DIFFERENTIAL PRIVACY

The MIT Press Essential Knowledge Series

A complete list of books in this series can be found online at https://mitpress.mit.edu/books/series/mit-press-essential-knowledge-series.

DIFFERENTIAL PRIVACY

SIMSON L. GARFINKEL

The MIT Press | Cambridge, Massachusetts | London, England

© 2025 Massachusetts Institute of Technology

This work is subject to a Creative Commons CC-BY-NC-ND license.

This license applies only to the work in full and not to any components included with permission. Subject to such license, all rights are reserved. No part of this book may be used to train artificial intelligence systems without permission in writing from the MIT Press.

This book was made possible by generous funding from the Alfred P. Sloan Foundation.

Illustrations by Ted Rall.

The MIT Press would like to thank the anonymous peer reviewers who provided comments on drafts of this book. The generous work of academic experts is essential for establishing the authority and quality of our publications. We acknowledge with gratitude the contributions of these otherwise uncredited readers.

This book was set in Chaparral Pro by New Best-set Typesetters Ltd. Printed and bound in the United States of America.

Library of Congress Cataloging-in-Publication Data

Names: Garfinkel, Simson, author.
Title: Differential privacy / Simson L. Garfinkel.
Description: Cambridge, Massachusetts : The MIT Press, [2025] | Series: The MIT Press essential knowledge series | Includes bibliographical references and index.
Identifiers: LCCN 2024019388 (print) | LCCN 2024019389 (ebook) | ISBN 9780262551656 (paperback) | ISBN 9780262382168 (pdf) | ISBN 9780262382175 (epub)
Subjects: LCSH: Privacy-preserving techniques (Computer science)
Classification: LCC QA76.9.P735 G37 2025 (print) | LCC QA76.9.P735 (ebook) | DDC 005.8—dc23/eng20240924
LC record available at https://lccn.loc.gov/2024019388
LC ebook record available at https://lccn.loc.gov/2024019389

10 9 8 7 6 5 4 3 2 1

To John Abowd and Christa Jones, for their trust and support that brought me to differential privacy and the 2020 Census.

CONTENTS

SERIES FOREWORD

The MIT Press Essential Knowledge series offers accessible, concise, beautifully produced pocket-size books on topics of current interest. Written by leading thinkers, the books in this series deliver expert overviews of subjects that range from the cultural and the historical to the scientific and the technical.

In today's era of instant information gratification, we have ready access to opinions, rationalizations, and superficial descriptions. Much harder to come by is the foundational knowledge that informs a principled understanding of the world. Essential Knowledge books fill that need. Synthesizing specialized subject matter for nonspecialists and engaging critical topics through fundamentals, each of these compact volumes offers readers a point of access to complex ideas.

PREFACE

I first started hearing about differential privacy in 2010, but didn't pay close attention. Frankly, I was intimidated by the math.

Differential privacy (sometimes abbreviated DP, but generally not capitalized unless in a title or at the start of a sentence) comes from a branch of computer science inhabited by people who are exceptionally adept at math and eager to share their excitement with others. They do this with proofs and definitions, the universal language of mathematics. Unfortunately for me, such explanations typically have symbols and terminology that I find unfamiliar and confusing. You see, I haven't taken a math course since high school.

Practically every paper and book about DP (other than this one) begins by reciting DP's mathematical definition. It's as if DP's promoters think that if they can just explain the math in terms that are clear enough, readers will grasp the correctness and universality of the concepts that they are about to present.

I am not a mathematician. I also learned the hard way during my second year in college that I am not a good chemist, so I switched my major to political science, got a master's degree in journalism, and spent much of the

1990s writing about digital security and privacy as a freelance writer. I returned to MIT in 2002 to pursue a midcareer PhD and then landed an amazing job as an associate professor at the Naval Postgraduate School in Monterey, California, where I developed techniques for finding hidden information in computers, cell phones, and inside digital files. There was no math!

I left the Naval Postgraduate School in 2015, and joined the National Institute of Standards and Technology (NIST) to work full-time on technical issues involving privacy. The US government was increasingly concerned about leaks of personal information—leaks from both hackers and stolen laptops as well as inadvertent leaks when documents or datasets were published on the web. At the time there was a growing interest in DP, but it wasn't nearly as flexible as other approaches for protecting privacy—approaches that had been used for years. Applying DP proved difficult. DP advocates responded that if an organization was serious about protecting privacy, there really was no other choice: the difficulties had to be embraced or else privacy might not really be protected. They had both the mathematics on their side to prove it and a growing number of cases in which traditional privacy-protecting techniques had spectacularly failed.

It turns out that's the big advantage of using math: once something is mathematically *proven*, it is unflinchingly, unambiguously true. There is simply no way to refute a correct mathematical proof. But I still wasn't convinced that the math was *necessary*.

The following year, I attended a workshop at Harvard University on data sharing. I recognized Dr. Kobbi Nissim, one of the inventors of differential privacy, at a lunch between sessions and walked over with my lunch tray in hand.

"I don't believe in differential privacy," I said to Nissim.

Perhaps it wasn't the best opening.

"What do you mean?" he asked. "You don't believe the math? That's like saying that you don't believe in multiplication."

Over lunch, I told Kobbi (as he likes to be called) that I thought DP was a neat theory, but it didn't seem practical to me. I couldn't use it to solve the real-world problems that I was dealing with at NIST, like how to de-identify the video clips from police body cams so that they can be released to the public without compromising the privacy of bystanders. In response, he said that de-identifying video is a poorly formed problem and not mathematically rigorous.

In conversations that I've had since with many DP researchers and practitioners, I've come to understand that the value of differential privacy is the mathematical certainty that it provides. This contrasts with other approaches for data privacy, which are based on best-guess efforts, but don't have any mathematical assurances that they actually work. Hence those failures that I mentioned above—and that I documented in NIST interagency report 8053, *De-Identification of Personal Information*.[1]

Consider police body cams: it's common to "de-identify" their video by blurring the faces of bystanders. But the bystanders may be standing next to their front door showing their apartment's street address. Using this kind of *auxiliary information*, it's possible to identify the people despite their blurred faces. This is the result of something called *the mosaic effect*, which I reference repeatedly in this book.

DP's mathematically rigorous approach removes the uncertainty of the mosaic effect: the math gives us worst-case bounds on what can be done with a specific batch of privacy-protected data. This certainty can be hugely comforting to leaders in government and industry.

On the other hand, because DP's bounds are "worst case," DP's certainty frequently results in more protection than is necessary. Because more privacy protection translates to data that are less accurate, this means that data protected with DP are sometimes less useful than they might otherwise be. Properly tuning this trade-off between data privacy and data utility is one of the fundamental challenges of using DP, but the flexibility to make that choice is one of DP's great advantages.

There are also many kinds of data for which there is no good way to apply DP (like police body camera videos) because the data modality doesn't fit well into DP's data model—at least not at the present time.

In January 2017, I moved from NIST to the US Census Bureau to work on the data protection mechanism for the 2020 Census. At that time, it was the most complex, consequential deployment of differential privacy that had ever been contemplated.

The data for the US decennial census are collected *under a pledge of confidentiality* and the legal protections of the US government's Title 13, both of which state that the Census Bureau will not publish or release any data that could be traced back to a specific individual or establishment. My job was to be the lead computer scientist overseeing the transition from legacy data protection methods to new methods based on DP.

At the Census Bureau, I had the opportunity to work with a team of amazing statisticians, mathematicians, geographers, demographers, and more, all furiously rushing to upgrade the privacy protection system for the decennial census. That effort received near universal support from the computer science community, but it was repeatedly attacked by a group of demographers and social scientists, who claimed that DP would make the census data less useful for their academic research.

In fact, DP ultimately created more opportunities for academic researchers by letting the Census Bureau release far more of the 2020 census data, and with more detail, than it had ever been able to release before. It also created significant interest, together with funding and research

Figure 1 Arguments in favor of differential privacy.

Figure 2 More arguments in favor of differential privacy.

opportunities, in the once arcane area of statistical disclosure limitation.

The use of DP was the subject of a federal lawsuit too. In 2021, the state of Alabama, Congressman Robert Aderholt, and two Alabama residents filed suit against the Department of Commerce, arguing that federal law did not authorize the Census Bureau to use DP.[2] Sixteen other states entered briefs in support of the suit after it was filed. Ultimately the suit was thrown out, although no court ever actually ruled on the legitimacy of DP.

The breakneck pace of the 2020 Census prevented those of us involved with the differential privacy implementation from deeply engaging with our critics. Many seemed opposed to the very idea of adding noise to the official statistics for the purpose of protecting privacy—with some critics apparently unaware that the Census Bureau had been adding various kinds of noise for decades. Other critics objected to the idea of differential privacy specifically.

To paraphrase Kobbi, I came away from these arguments thinking that many of DP's critics simply didn't believe the math. Although many critics acknowledged that there is a fundamental trade-off between releasing accurate statistics and protecting data collected under a pledge of confidentiality, they typically contended that DP inflicts unacceptable damage on the utility of data used for delivering public services, redistricting, and academic research—this, despite the fact that the amount of noise that DP adds is tunable. They also maintained that while legacy approaches developed for statistical disclosure limitation lack DP's *mathiness*, those legacy approaches may do a better job balancing privacy protection and data utility.[3]

Writing in the *Harvard Data Science Review*, the highly respected academics Daniel L. Oberski and Frauke Kreuter present a view somewhere in the middle:

> The main disadvantage of ensuring differential privacy is that it typically requires more noise infusion than traditional techniques. This is a consequence of the fact that traditional techniques only need to prevent linkage, while differential privacy prevents linkage through reconstruction. One might expect that in the discussion on how and when differential privacy should be applied, level-headed experts convene to weigh such pros and cons and find a consensus.
>
> But where we might expect a dry weighing of facts, we observe a heated debate, which shows no signs of abating (see Ruggles, Fitch, Magnuson, & Schroeder, 2019; and the 2019 Harvard Data Science Initiative Conference). Why do ordinarily equanimous researchers embroil themselves in

such a raging controversy? And, most puzzling, why hasn't evidence and scientific argument been able to adjudicate this apparently scientific disagreement yet?

We believe scientific facts have not been able to end the disagreement, because the disagreement is not about facts. Rather, the parties have different subjective beliefs about risk, and therefore differ in their ideas on how to mitigate such risks. Furthermore, the current debate often overlooks the social science behind data collection and privacy perceptions.[4]

In this book, I'll discuss some of these assertions and seriously consider the position of DP's critics. In the conclusion, I'll present possible compromises between legacy approaches for statistical confidentiality protection and a "pure-DP" future.

This book is written for a general audience, not for mathematicians, statisticians, or computer scientists. It provides a general understanding of what DP is, how it works, and why it is both necessary and revolutionary. You'll also learn about the people behind DP. This book does not, however, supply detailed step-by-step information for releasing datasets using DP and what to with the many kinds of data releases for which DP is not yet sufficiently developed. For readers in search of immediate help to such problems, I recommend the recently published NIST special publication 800–188, *De-Identifying Government Datasets: Techniques and Governance* as well as the article "Differential Privacy for Databases."[5] Programmers interested in adding DP to existing systems should look at the book *Programming Differential Privacy* and the OpenDP user guide.[6]

INTRODUCTION

DP is a mathematical framework for discovering, describing, quantifying, and ultimately controlling the maximum possible privacy loss that you might personally suffer as the result of an organization using your confidential data.

It was invented to protect personal data used for making statistics, such as the block-by-block redistricting files created as part of the 2020 US Census, but it has found many other applications. DP's success drives from its flexibility: the same mathematical framework that works to protect census data has been adopted to protect training data used for artificial intelligence, prevent internet advertisers from learning sensitive information about people seeing their ads, and even prevent overfitting in statistical analyses.[1]

While other statistical approaches for protecting privacy never saw much interest beyond a small community of researchers and practitioners, DP has heralded a revolution in mathematically formal approaches for protecting privacy that has been widely embraced not just in academia but also in government and industry. This is differential privacy's lasting achievement.

DP protects confidential data by adding in random numbers from a carefully calibrated statistical distribution. This noise makes it arbitrarily difficult for someone to take a published statistic—like the ages and races of people living on a census block—and learn the underlying confidential data on which the statistic is based. DP's reliance on formal mathematics means that anyone with an understanding of DP can check the math and verify the protection that a DP system provides; DP requires no reliance on expert opinions.

Differential privacy was invented by Cynthia Dwork, Frank McSherry, Kobbi Nissim, and Adam Smith, and announced to the world in their 2006 paper, "Calibrating Noise to Sensitivity in Private Data Analysis."[2] These computer scientists introduced a fundamentally new approach for protecting privacy—one we now call *formal privacy*. All the previous approaches were based on a kind of "arms race" between defenders and attackers. A defender would come up with a new approach for protecting personal information, and an attacker would then show that the approach didn't actually work. Frequently these attacks were discovered after large datasets of supposedly safe information had published for researchers on the internet.

DP has heralded a revolution in mathematically formal approaches for protecting privacy that has been widely embraced not just in academia but also in government and industry. This is differential privacy's lasting achievement.

Figure 3 Differential privacy's four founders created a fundamentally new way to think about privacy.

DP's approach is based on a mathematical definition for a broad class of algorithms that take personal data and produce results that are useful as well as privacy protecting. Algorithms that follow this definition automatically enjoy all the benefits of differential privacy. Thus it's relatively straightforward for a mathematician to formally evaluate a proposed privacy-protecting approach and determine if it does or does not meet DP's requirements. Experts cannot disagree: the proofs either work or they do not.

It's relatively straightforward for a mathematician to formally evaluate a proposed privacy-protecting approach and say if it does or does not meet DP's requirements. Experts cannot disagree: the proofs either work or they do not.

Figure 4 The transparency paradox.

As a result, DP does not require having an expert decide which data in the database are "personally identifiable" and which are not; it protects every bit of confidential data. Likewise, DP makes no assumptions about the technical capabilities, computing power, or even auxiliary data that a potential attacker might have. DP does not require that algorithms, source code, or parameters be kept secret; in fact, DP practitioners encourage the release of such information in the interest of transparency. For these reasons, DP is frequently called the "gold standard" of privacy protection technology.

Protection against the Mosaic Effect

DP protects every person in the database—people in the middle of the distribution as well as the outliers. It requires that every use of each piece of data coming from every person is accounted for—and recognizes that every data use causes some privacy loss. As a result, DP forces organizations to make principled trade-offs between data use and privacy protection. *Principled* means that the organizations are forced to adjust the trade-off between the accuracy of the statistical products and the resulting privacy loss to the data subjects. This is at the root of why some deployments of DP have been controversial.

Moreover, DP is the only data privacy approach today that takes into account the possibility of combining information from multiple data releases to learn information that was supposed to be kept confidential—something called the mosaic effect, as mentioned earlier.

We are all familiar with the mosaic effect, although not necessarily by that name. Fictional detective Sherlock Holmes called it the "Science of Deduction and Analysis." Consider a child who comes home from school every Friday with a big grin on their face and an A+ on their graded homework. Two months into the school year, the child comes home upset and without their homework. Meanwhile, the parent sees a Facebook post from another parent about taking another child out for ice cream to celebrate a good grade on the latest assignment. When the child's parent infers that there was in fact an assignment and the child in question got less than an A+, that's the mosaic effect at work.

Here's another example. In 2021, the *Pillar*, a Catholic publication, published a report detailing how a specific church official had been visiting gay bars and using "location-based hookup apps in numerous cities from 2010 to 2020, even when traveling on assignment for the U.S. bishops' conference."[3] The *Pillar* performed its analysis with commercially available datasets, according to press accounts. The datasets did not contain names, but did contain device identifiers, location information, and information about specific cell phone applications. Because the residences, offices, and travel plans of high-ranking clergy are easily discovered, it was straightforward to sift through the data and learn the identity of the person using the hookup apps and presumably betraying his vows.[4]

The mosaic effect is a concern for government agencies that publish statistics based on confidential data. These agencies are legally required to ensure the secrecy of the data they collect under a *pledge of confidentiality* (also called an *oath of nondisclosure* or *sworn affidavit of nondisclosure*). But they are required to publish useful statistics too. So these agencies need to ensure it is not possible to combine information from different releases and then deduce the underlying sensitive information. Unable to repeal the laws of mathematics, these agencies face a quandary: they must trade-off the damage to individual privacy with the public benefit of the published statistics.

Protecting confidentiality is especially difficult given the push for agencies to publish *microdata*, a term commonly used to refer to the individual data records about people, establishments, or some other kind of "unit" that are collected, edited, tabulated, and otherwise used for statistics making. With microdata, researchers outside the statistical agencies can perform their own analyses and even combine data from multiple agencies. In practice, agencies remove the names and other highly identifying attributes before they publish microdata. The danger of publishing microdata is that it is sometimes possible to reidentify the de-identified data. The mosaic effect makes it much easier to mount this kind of attack on public microdata. The world of official statistics uses the term *data intruder* to refer to a person who performs such an attack, but I prefer the term *data hacker*.

Confidential Data for the Public Good

Differential privacy marks the most important discovery to date in the quest to protect confidential data from individuals while making use of their data for the public good. Many people trace the formal articulation of this research

Figure 5 Census workers are pledged to secrecy, but what happens to the data after they are collected is a matter of data access policies and the effectiveness of statistical disclosure limitation techniques.

problem to a 1977 article by the great Swedish statistician Tore Dalenius (1917–2002).[5] Dalenius's core hope was to develop an approach for using confidential data such that nothing could be learned about any specific individual other than what could be learned about the group as a whole through published aggregated statistics.

Dalenius's desiderata have far-reaching consequences for today's data-rich society:

- We use educational statistics about class sizes and grade averages to help parents decide where to send their children to school—and allocate substantial amounts of taxpayer dollars. Is it possible to create, publish, and use such statistics without revealing the educational records of individual students?

- Our society spends billions of dollars subsidizing housing and transit. Increasingly, maintaining public support for these programs requires being able to generate their effectiveness, which means taking the data of the specific people involved in these programs and demonstrating how they benefited, and how that benefit improved society. Is it possible to produce such evidence-based analyses without revealing information about those specific individuals?

- Many of the information services available on today's internet are paid for by advertising. We know that the most effective advertising is targeted using private information about the person who is then shown the advertisement. At the same time, laws increasingly prevent the sharing of private information with advertisers. Is it possible for an advertising network to display the targeted ad while keeping that private information confidential?

Dalenius wanted questions like these to have the answer "yes," and proposed a set of rules that made it possible to publish tables of statistics without directly revealing information about individuals. But these rules break down when there is additional information about individuals in the dataset; the rules don't take into account the mosaic effect.

Privacy versus Confidentiality

The word *differential* in the phrase *differential privacy* is designed to evoke "the increased risk to one's privacy incurred by participating in" a data collection effort and having the resulting data placed in a database that is used for creating a statistical product, compared to that same statistical product being created without your data being present.[6] The intuition is that an analysis of a database that does not have your data cannot violate your privacy.

This is one of the breakthrough achievements of the 2006 DP paper: a useful and mathematically rigorous definition of privacy that does not depend on the presence (or absence) of external information. Previous efforts to include these kinds of mosaic effects into approaches for protecting privacy proved unworkable; protection mechanisms that appeared to work in all cases suddenly failed when new datasets were discovered. This created real-world problems for organizations like the US Census Bureau, which previously needed to evaluate what data might already be publicly available before releasing new microdata.

DP solves this problem by focusing entirely on the differential impact of a data release based on your confidential data compared to a data release that did not use your confidential data. Using this intuitive leap alone, it is possible to derive the mathematical equation for differential privacy—although please don't ask me to do it!

But what about that word *privacy*?

Today, one of the primary meanings of privacy is the ability of individuals to control information about themselves. This includes each person's decision of whether to provide information when requested and the ability to control how that information will be used. Somewhat confusingly, we also use the word to describe if a person can assure that they are free from surveillance and intrusion.

Professor Daniel Solove at the George Washington Law School is a leading thinker on privacy. In 2006, Solove created a "taxonomy of privacy" to help bring order to

Table 1 Daniel Solove's Taxonomy of Privacy

Activity group	Harmful activity	Example
A. Information collection	1. Surveillance	Your neighbor across the street uses a camera to monitor your front door.
	2. Interrogation	Your neighbor asks you about a person they saw entering your residence.
B. Information processing	1. Aggregation	Analyzing the front door photos, your neighbor creates a log of every outfit you have worn for the past two years.
	2. Identification	Using DNA collected from your front door, your neighbor searches your true identity with an online consumer genetic service.
	3. Insecurity	Your neighbor does not protect the information about you that was collected and now many people have access to it.
	4. Secondary use	You asked your neighbor for the photos of you and your family members in their database. Your neighbor needs photos to match against, so you supply a headshot for each person to match against. Your neighbor takes these headshots and sells them.
	5. Exclusion	Your neighbor collects information about you, but you never learn about their secret database.

Table 1 (continued)

Activity group	Harmful activity	Example
C. Information dissemination	1. Breach of confidentiality	Your neighbor promises that the database about your comings and goings will be kept confidential, but it's released.
	2. Disclosure	**The existence of the database itself and fact that it contains your photos is disclosed to others.**
	3. Exposure	Included in the database was a photograph of you, naked, taken through your window.
	4. Increased accessibility	Your neighbor has gone to your town's library, checked out copies of your high school yearbook, and added all the photos of you to a public website.
	5. Blackmail	Your neighbor's camera records you engaged in an illegal act, and your neighbor then threatens to report your activity to the police unless you pay them $10,000.
	6. Appropriation	Your neighbor takes a photo of you and your family holding a cake that you have baked and uses it in an advertisement for a local flour mill. The photo runs under the headline "Flour of the Family."
	7. Distortion	The database that your neighbor creates and published says you left home at 9:30 a.m. on many days that you know you left at 8:00 a.m. and arrived at work by 8:30 a.m.

Table 1 (continued)

Activity group	Harmful activity	Example
D. Invasion	1. Intrusion	Your neighbor wants better photos, so they pick the lock on your front door and enter your house.
	2. Decisional interference	You know that your neighbor is watching what you do, so you change your behavior and ultimately decide to sell your house.

Notes: The harmful activity in **bold** (disclosure) is the primary activity that can be mitigated using differential privacy; it involve the creation of a database in which the individual data records, search terms, or query results can be protected through the addition of noise. It is possible, however, to imagine ways that DP might limit the ability to find individuals in a DNA database (identification) or limit additional use (secondary use) of data. *Source:* Daniel Solove, "A Taxonomy of Privacy," *University of Pennsylvania Law Review* 154, no. 3 (January 1, 2006): 477–560, https://scholarship.law.upenn.edu/penn_law_review/vol154/iss3/1.

scholarly discussions about privacy. By reviewing decades of court cases, legislation, and the writings of other scholars, Solove came up with sixteen distinct kinds of harmful activities that resulted in people feeling that their privacy has been invalidated; he placed these activities into four distinct groups. The full list appears along with my examples of the activities harmful to privacy in table 1.

Differential privacy is designed to help limit the harms that result when confidential information is used to produce a publication like a report, single statistical table, or dataset that then impacts the outside world. This impact might be something as consequential as redistricting the US House of Representatives or making data for the billion-dollar federal student aid program available over the web. Or it might be something relatively inconsequential, such as publishing statistics in a stodgy economics paper, showing an advertisement to a person over the web, or even selecting which song to play over a person's streaming music service.

That is, the purpose of differential privacy is to protect the *confidentiality* of data. Using confidential data inherently reveals information. The policy question that differential privacy forces us to confront is this: *How much information should be revealed, and for what benefit?*

Figure 6 shows the basic model we'll use for discussing this question.

Transparency and Kerckhoffs's Principle

Another aspect of differential privacy is its transparency, a notion that comes from DP's roots in computer security and cryptography. The deep idea is that the best way to create a secure system is to use open design principles; security should come from the strength of the design rather than from its secrecy.

This principle was first formulated by Dutch cryptographer Auguste Kerckhoffs in 1883. One of the best examples of its use today is the development and implementation of the Advanced Encryption Standard by the

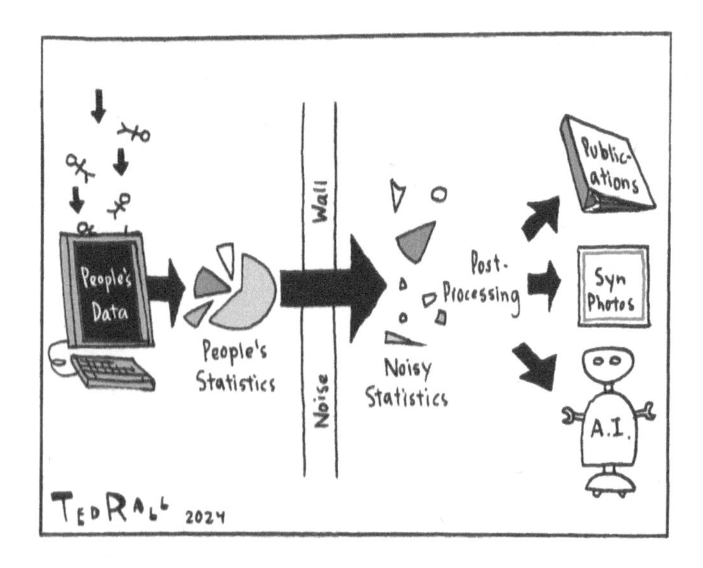

Figure 6 In the basic model of differential privacy, a trusted curator collects confidential data from people, stores that data in a database, uses the data to create statistics, applies statistical noise to the statistics, and finally publishes the results. The "noise wall" separates the confidential data on the left from the protected data on the right. The privacy guarantees of differential privacy are immune to postprocessing, which means that any operation performed on the data to the right of the noise wall cannot result in additional privacy loss. This contrasts with other privacy mechanisms, where multiple protected datasets can sometimes be combined to learn new privacy-invading details.

National Institute of Standards and Technology, which involved hundreds of researchers working at dozens of research institutions throughout the world in the 1990s and produced what is widely regarded even now as one of the world's strongest encryption algorithms.[7]

Although Kerckhoffs developed his principle for cryptography, today it's commonly applied to security engineering in general with the slogan "no security through obscurity." Differential privacy applies this principle to statistical privacy protection.

Applying Kerckhoffs's principle to statistical privacy, the goal is to release statistics without relying on confidentiality mechanisms that if known, could be readily undone. Open design principles make it possible for anyone to evaluate the system, probe for weaknesses, and offer ways to improve it.

Confidentiality in Statistical Databases

DP is a key development in a long line of research on the problem of providing confidentiality for information stored in *statistical databases*. Such a database might be a collection of health records, tax returns, or even the list of links that a person clicks on the internet. It is a problem that was studied for decades with little real progress because of the mosaic effect.

Consider a school with a database of student grades. Researchers might want to know if girls and boys in single-sex classrooms offered by some schools get better grades on average than girls and boys in mixed-gender classrooms. A statistical database that includes every student's demographic information, every course each student took, and the grades that each student received for every assignment might help researchers answer this question. The researchers could directly compare the grades of students in single- and mixed-sex classes. They could also consider the grades of students who switched between single- and mixed-sex classes during their time in school.

The researchers would formulate their questions, submit them to the person running the database, take the results, and finally write their paper. In the world of differential privacy, each of these questions is called a *query*, and the person operating the database is known as the *trusted curator*.

To protect student confidentiality, the curator needs to block queries that might reveal student names or grades that could be easily traced to a single student. This isn't just good ethics; in the United States, this is a requirement of the Family Educational Rights and Privacy Act (FERPA). It's a requirement that's more complicated than it seems! For example, researchers might ask, "How did the grades of girls change for the girls who transitioned from mixed-sex fifth grade math classes to single-sex sixth grade classes?"

Answering such a query might violate FERPA if only a single girl made such a transition. But programming databases to avoid every possible way that a query could violate a person's privacy turns out to be a much harder task than it seems at first.

Database Reconstruction

As computerized statistical databases became more widespread in the1960s, database developers proposed a variety of commonsense approaches to protect the confidential data that such databases contained. One early idea was to program the databases so that they would only release statistics based on more than a minimum number of students. To use our example above, the database might be programmed to answer a query only if it was based on data from more than five students. So the query above would be answered if five girls transitioned from mixed- to single-sex classes, but not if only a single girl transitioned.

Yet for every privacy-preserving approach that was proposed, computer scientists found ways to circumvent the protections by asking a series of carefully chosen queries.[8] For instance, if ten girls transitioned, but nine of those girls were also in a single-sex English class, a person bent on defeating the privacy mechanism could simply ask two queries. The first query would be for all ten girls, and the second would be for the girls who were also in the English class. The researcher could then subtract the two results; the difference would be based solely on the one student who was not in the single-sex English class.

This approach of combining the results of many queries to recover specific details of an underlying confidential database is called *database reconstruction*. It's easier with auxiliary information, but such information is not required, as demonstrated by popular games like Mastermind and Wordle. In these games, the confidential "database" is not the list of all possible patterns or words but rather in the specific hidden value that is being queried. A Wordle player relies on auxiliary information as well: the knowledge of which combination of letters make up valid words.

Here is a more complex database reconstruction problem:

> You are in a math class with nine other students. Your teacher hands back your midterm exam and then congratulates the class for getting a class average of 98. You peek at your exam; you got an 80. Give the scores of the other nine students.

The answer depends on whether or not your teacher gives extra credit. If the top possible grade was 100, then that is (not coincidentally) the grade that the other nine students received. That's because the average of nine perfect scores and 80 is 98. But if there were five possible extra credit points, then there are many other arrangements of integer grades where one grade is 80 that have the average of 98. For example, if you have an 80 and another student could have a 60, then the eight other students would all have 105. Or there might be three students with 99, three students with 100, and three students with 101; your 80 brings the class average down to 98. Each of these combinations is a possible reconstruction. In this case, where grades are integers between 0 and 105 (inclusive), and where one grade is 80 and the average is 98, there are 14,287 reconstructions possible.

Let's stick to the case where there was no extra credit: you got an 80, and all of your classmates got 100. If those nine students are all friends, they probably shared their perfect scores with each other. By colluding in this manner, those students have now learned your grade. By releasing the class average, your teacher inadvertently disclosed your grade and likely violated FERPA.

Statistical Disclosure Limitation

There are several approaches that your teacher could have used to prevent your classmates from learning your grade:

- **Suppression**: Your teacher could simply not report the class average.

- **Generalization**: Instead of saying that the class average was 98, your teacher could have said that the class average was between 95 and 100.

• **Noise infusion** (also known as noise "injection"): Your teacher could have said that the class average was 96.5 after adding a small random number. (The noise can be added either to each grade or the final static.) Such random numbers are called *noise*.

Each of these approaches has a problem:

• *Suppression* seems to do a good job protecting confidentiality, but it also prevents the other students from getting a sense of how well they did. Furthermore, it potentially reveals information: if your teacher normally releases grades but this time didn't, then you know that this time your classmates all got perfect grades!

• In this case, *generalization* would reveal to the other nine students that you scored somewhere between 50 and 100.

• *Noise infusion* is perhaps the strangest idea of the three. How frustrating! Your teacher seems to be releasing intentionally bogus statistics!

"How big was the random number?" one of those super-smart kids in the front of the class might ask.

"I can't tell you, but riddle me this," says the teacher:

In the hall of games there is a table with dice,

and one of those dice has 365 sides.

When the grading is done, I go to the hall.

I find the die and roll for your toll.

If the day on the calendar doesn't match the day on the side,

I add a number between −5 and 5, and hide.

A smile breaks out across your teacher's lips.

"And if they match?"

Your strange math teacher cackles, "In that case, I add or subtract a bigger number!"

For readers who find this riddle a bit too riddlesome, suffice it to say that the teacher will add or subtract random number between 0 and 5 with 99.73% probability, and a larger number with 0.27% probability.

Suppression, generalization, and noise infusion are all statistical disclosure limitation techniques. The term *disclosure* is shorthand for "disclosure of confidential data on which the statistics are based." The word *limitation* means that we can limit the amount of disclosure that results from publishing statistics, but we can't eliminate it.

Statisticians sometimes use the phrase *statistical disclosure control* to describe these tools. I prefer the term *statistical disclosure limitation* because the word *control* implies that statistical disclosure can be eliminated, whereas the math tells us that it can only be reduced.

Many texts that cover statistical disclosure limitation say that there is no one best control and therefore the choice of which control to use is best made by an expert in the field. The radical premise of this book is that such guidance has been rendered obsolete by advances in mathematics. We now know that adding uncertainty to the published statistics is always the best choice for protecting privacy. This book also argues that the decision of how much noise to add is a policy decision because there is no "right" amount of noise to add; it is always a political trade-off.

Differential privacy is a form of noise infusion, where the noise is precisely crafted according to rigorous mathematical rules.

In this toy example, it may seem that generalization is a better choice for the students in the class, especially given that a range consisting of the top 5 percent of possible averages (95–100) has a median of 97.5, and this median is close to the ground truth value (98). But a significant problem with generalization is that it's hard to combine multiple releases that use generalization in a way that is guaranteed to preserve privacy. To use the language of differential privacy, generalization is not *composable*.

The Composability Advantage

Critics of differential privacy frequently argue that DP is far more complicated than is necessary—that traditional statistical disclosure limitation techniques overseen by experts in that technique offer practical privacy protection without the hassle of DP's math, and without the uncertainty of noise and random numbers. This argument might seem attractive when evaluating a single data release, but it falls apart when the same confidential data are used multiple times for many releases or there are external facts that can be correlated with a release that is intended to protect privacy. The assertion also fails to take into account the mosaic effect.

DP avoids the mosaic effect because its privacy protection is *composable*—that is, the math takes into account the fact that a data hacker might combine multiple DP releases or combine those releases with other public data.

Computer scientists use the word *composable* to describe a system design property that small modules and systems behave more or less the same when combined together to create larger systems. A handheld calculator is composable because you can take a complex calculation and run it on a single calculator, or split it up into pieces and run parts of the calculation on many different calculators, and then combine the results.

Traditional techniques for statistical disclosure limitation are not composable. Here's an instance of why that can be a problem. Let's say the school wants to publish a report of its student grades, but it doesn't want to use differential privacy. One report might say that all students in grade 9 math had a class average between 80 and 90.

The school is proud of this; it notes that last year's grade 9 math had an average between 75 and 85.

Now suppose the school produces a report based on a troubling statistical finding: students from neighborhoods to the east of the school tend to have lower scores than the students who come from neighborhoods to the west. Half of the students from the east have class averages less than 60, the report notes, a quarter have averages between 61 and 70 (inclusive), and another quarter have scores that are between 71 and 80. Not a single student living east of the school has a score higher than 80.

You know what's coming next: there is one student in grade 9 math who comes from the east. You are that student. You have a single score that's been sorted into two different sets of buckets as part of the two different data releases. Nobody knew that this was the case because both releases were produced using de-identified data—meaning none of the scores had your name on them. Everybody involved thought your privacy was being protected. But because de-identification and generalization don't mathematically compose, the privacy desiderata of the two statistics are undone when they are combined. That is, your school has published data from which anyone can mathematically deduce that there is a single student from the east who has a grade 9 math score of 80—and that's you.

Adding noise to the reported statistics would overcome this composability problem. If each statistic were published with a small amount of noise, an apparent overlap might be the result of single student who just happened to be in both datasets or instead simply noise. This uncertainty is what powers DP's privacy protection.

Differential Privacy's Key Insight

The key intuition behind differential privacy is that your privacy is not impacted by someone who publishes an analysis on a dataset that doesn't contain your data. This idea may seem obvious, but it is deep.

Let's go back to the math class. How much variation could your score of 80 possibly have had on the result that your teacher revealed? Let's explore two counterfactual possibilities:

World A1: In this world, your score was 0 (rather than 80). The true average would be 90.

World A2: In this world, you got 100, just like everyone else in the class. Now the true average is also 100, because everybody got a perfect score.

These two worlds represent the extremes that you could have scored on the exam and show how much your experience could have impacted the true average. In DP, this is called computing the *sensitivity*. Given that your possible score could range 100 points, the most that your score could impact the reported average is 10 points. This computation does not depend on the actual grades that you and your classmates received; it is inherent in the design of this system. So if your teacher reported the class average as "somewhere between 90 and 100, inclusive," there is no way that the other nine students could infer your score, even if they all exchanged their grades over lunch while you were doing something else.

Unfortunately, a report that the class average was somewhere between a B- and an A+ isn't useful for anyone trying to use the data. It also doesn't fully protect the privacy of the other nine students because it places tight bounds on their minimum and maximum possible grades. And if the school is in the habit of releasing the true range rather than predetermined range buckets, it still reveals your score of 80 to the other students in your class!

With differential privacy, the data curator reports the statistic based on confidential data as a single number. In the example above, the statistic is the class average. Your teacher reports that the average is 96.5, but we as sophisticated data users know that (because of the noise) the grade has a 95 percent probability of being somewhere between 94.5 and 98.5.

That's about as far as we can go with pure DP. The are variants of DP, however, such as *approximate differential privacy* and *zero concentrated differential privacy* (zCDP), both of which significantly reduce the amount of noise that DP adds for similar levels of privacy protection. I'll discuss those variants later in this book.

Differential Privacy's Brief History

Dalenius's goal of statistical disclosure limitation was to allow for the publication of statistical data without the possibility of revealing information about the individual people or establishments on whom the data are based.[9]

The *fundamental law of information recovery*, discovered by Irit Dinur and Kobbi Nissim in 2003, proves that Dalenius's goal is mathematically impossible: *every* statistical release based on confidential data leaks some aspect of the confidential data on which it is based.[10] If the results of enough queries are released, the entire confidential database is eventually revealed.

While he was working on that 2003 paper, Nissim visited Cynthia Dwork at Microsoft Research (MSR) in California. Microsoft had acquired a consumer email system called Hotmail in 1997 for $500 million, and Dwork was researching ways that Microsoft could use the data in such internet-scale datasets while preserving the privacy of the people who had provided the data.

Nissim and Dwork agreed that the attack that Dinur and Nissim were developing in the 2003 paper required too many queries to be practical, so Dwork and Nissim starting exploring what would happen if there were fewer queries. The result of those discussions appears in the fourth section of Dinur and Nissim's 2003 paper, which the paper's acknowledgments state "is joint with Cynthia Dwork."[11]

Nissim moved to MSR in 2003 and started working on formal privacy with Dwork. Their next step was a 2004 paper on privacy-preserving data mining.[12] The following summer, Avrim Blum, Dwork, Frank McSherry, and Nissim (who by then had taken a faculty position at Ben-Gurion University) wrote a paper that examined the power of using the combination of noise and sums ("noisy sum queries") as a computational building block and showed how to carry out a variety of data mining tasks.[13]

Dwork presented the key ideas at a July 2005 workshop in Bertinoro, Italy, that she organized with Stephen Fienberg (now deceased) of Carnegie Mellon University. In the workshop, Fienberg criticized the approach, reportedly saying, "Your utility is going to be in the toilet."[14] But the team was not deterred, and Dwork gave another talk on the idea at Cornell that fall to celebrate the computer science department's fortieth anniversary. At the time, the idea was called *incremental privacy*. The phrase *differential privacy* was suggested by Mike Schroeder at MSR shortly thereafter when Dwork and McSherry were preparing a presentation for Bill Gates. The name was better than the others that had been proposed and so *differential privacy* stuck.

That fall, Dwork gave more talks with the name "differential privacy" at the University of Maryland and University of Washington, while she worked with McSherry, Nissim, and Adam Smith, who had previously been an intern with Dwork, on the foundational DP paper, "Calibrating Noise to Sensitivity in Private Data Analysis."[15]

It is typical in the computer science research community for work to be presented by the most junior coauthor of the team, so Smith, then a postdoctoral researcher at the Weizmann Institute, presented the "Calibrating Noise" paper in March 2006 at the Theory of Cryptography Conference in New York City.

"I think the paper's reception was politely positive but not enthusiastic," Smith told me in the fall of 2023. "I think it wasn't clear to people whether this was just one more step along the way, or if that line of work had really 'arrived.'"

Ironically, the phrase *differential privacy* didn't appear in the 2006 foundational paper. Instead, Dwork used the name as the title of a plenary lecture she gave that July in Venice, Italy, at the thirty-third International Colloquium on Automata, Languages, and Programming.[16] Her lecture reported on these results and presented a formal argument that she had worked out with Moni Naor that it was formally impossible to achieve the goals outlined in Dalenius's 1977 article.[17] But because this second 2006 paper has the title "Differential Privacy," Dwork's solo paper is frequently mistaken as the paper that introduced DP.

Remarkably, the US Census Bureau adopted DP just two years later, in 2008, to protect the income and commuting patterns in the OnTheMap interactive tool that had been under development for several years.[18] OnTheMap was designed from the start to use noise infusion to protect privacy. With the discovery of DP, there was now a theoretical framework that the Census Bureau could use to determine what kind of noise to use, how much noise to use, and how to add it. This experience shows that it can be readily straightforward to use DP when noise infusion is designed into a statistical application from the beginning.

In 2014, Google deployed a version of differential privacy in the Chrome browser to collect information about each users' home page and the programs running on their computers.[19] Apple and Microsoft deployed DP in their operating systems in 2017 to collect details about, respectively, emoji choices (Apple) and operating system performance (Microsoft); that same year, Uber announced that it had developed a database that used differential privacy to protect customer information viewed by analysts.[20] In 2017, the US Census Bureau announced that it would likely be using DP to protect the 2020 census.[21] In 2018, it announced that the "End-to-End Test" would be protected with DP, and DP was formally added to the 2020 Census Operational Plan the following year.[22]

There are now a growing number of DP deployments in the United States and overseas. What was once a fringe idea from a group of theoretical computer scientists is changing the way the world computes and releases data based on confidential information.

The Privacy Loss Budget

Central to the operation of differential privacy is a parameter known as the privacy loss budget (PLB), which is the total privacy loss that an organization allocates to a particular statistical release. Setting the PLB lets statistical officials implement the trade-off between protecting privacy and making use of confidential data called for by various laws and public policy.

I recommend not viewing the PLB as a budget in a traditional sense; I haven't seen an actual DP deployment in which the PLB was authorized by a panel of wise people who have decided in advance what is best for individual privacy. Instead, I've seen the PLB used most often as an aspirational budget—the kind that a parent might give a ten-year-old child buying clothes for the first day of school.

Child: "Mom! I need new purple shoes!"

Mom: "Yes, you can spend an extra $48 on new shoes. Now you are spending $412 on clothes, which means we are turning off your streaming video service this fall."

The PLB works like that, except for confidentiality. Especially now, in the early years of differential privacy, PLB is something that organizations want to minimize because higher PLBs mean that they are making more use of confidential information, which likely means that individuals are being exposed to more risk. This in turn may expose the organization itself to more risk. Because we don't yet have a good way of translating privacy risk to financial or

Figure 7 Differential privacy makes it possible decide the trade-off between accuracy and privacy loss. In the early days of DP, some computer scientists argued that the PLB, represented by the Greek letter epsilon (ε), should have a value of not more than .01 in some cases.

reputational risk, the idea is to simply minimize PLB as much as possible. The PLB lets the organization implement a policy that give less accuracy to applications that are less important and drop others altogether on the grounds that such uses would substantially increase the PLB.

Child: "How about if I only get two new pairs of pants, and we'll keep the streaming service? And I don't really need the belt. Can I get two pairs of shoes instead?"

The PLB makes it possible for organizations to make these kinds of choices. Typically, there are many statistics being produced; the PLB makes possible a kind of privacy accounting, where the accuracy and associated privacy loss are meted out between competing interests.

Researcher: "I need to know the average GPA of the girls separately in their math and English classes, the boys in the same classes, the girls that transitioned from single-sex classes to mixed classes, the boys that transitioned, and then a subsample of all of those students who happen to be in the bottom quarter of family income."

Trusted curator: "That's a lot of statistics! Last year's analysis had a PLB of 8. If we keep the same level of accuracy, this year you're looking at a PLB of 14. We will be heavily criticized when we publish our annual report."

Researcher: "OK, let's not run the family income statistics; they will be less accurate because they will be based on fewer students. Let's just release the GPA for math. I wanted the other numbers for a new grant application that I'm writing, but I can use just the math numbers for the application."

Privacy, Confidentiality, and the US Census

In 2019, Dwork explained at the 2019 Harvard Data Science Review Inaugural Symposium in Cambridge, Massachusetts, that she and her colleagues had invented DP specifically to solve the kind of privacy problems encountered by the US decennial census—a program for which participation is mandatory and the results are publicly distributed.

The existential purpose of the decennial census, as specified in the US Constitution, is to determine the number of seats in the US House of Representatives that each state is allotted. This was part of the "Connecticut Compromise" worked out in Philadelphia during the sweltering summer of 1787.

The Constitution specified the number of representatives that each state would have for the first Congress (1789–1790), but required Congress to have a census in 1790 so that the House could be properly apportioned for the second. It's the decennial census that makes the House "representative" of the people. The Senate gives two seats to each state, ensuring senators represent the interest of their entire state—and indeed, US senators were appointed by state legislators, rather than directly elected by the people, until the passage of the Seventeenth Amendment to the US Constitution in 1913.

A *census* is a count of every individual or item in a specific population. The first-known census was in ancient Babylonia in 3800 BCE. For millennia, their primary purpose was to support conscription and taxation, as evidenced by the Lord's commandment in Exodus 30:12 to take a census of the Israelites and collect a "ransom" for each man. Sweden was the first nation to establish a regular national census in 1749, although Sweden's census was not used to draw legislative districts or otherwise apportion political power.[23] The idea of using a census to draw legislative districts is a US invention. The first US census was in 1790, the second in 1800, and so on. We haven't missed one yet, so the 2020 Census is also known as the twenty-fourth census. As I write this book, the Census Bureau has already started planning for the twenty-fifth census, which will take place in 2030.

The Constitution specified that both the number of state representatives and the direct taxes apportioned by the federal government to each state "shall be determined by adding to the whole Number of free Persons, including those bound to Service for a Term of Years, and excluding Indians not taxed, three fifths of all other Persons." So from the beginning, the decennial census was more than a simple head count; it had to collect details of each person.

Specifically, the 1790 Census had to distinguish between people who were free and the people who were in bondage. At the first Congress, Virginia representative James Madison and New Hampshire representative Samuel Livermore debated whether the first census should collect additional information. Madison wanted information to help the central government craft better laws and policies. Livermore was concerned that collecting more information would be more costly and inevitably result in higher taxes as the government learned about pockets of wealth in the countryside.[24]

Livermore's argument carried, and the 1790 Census collected just six pieces of information for each family:

- The name of the family's head
- The number of free white males age sixteen and older
- The number of free white males under sixteen
- The number of free white females

- The number of all other free persons
- The number of people who were in bondage[25]

Today we might be tempted to view Livermore's objection through the lens of privacy. Applying Solove's taxonomy from table 1 above, we might say that Livermore wanted to limit the collected data's potential for secondary use.

An analysis using DP's concept of privacy yields a different answer. Livermore wasn't concerned that the 1790 census records *for a specific resident* might result in a tax collector showing up at that person's doorstep and demanding unpaid taxes; he was worried that Congress would use the *statistical knowledge gleaned from the Census* to create new taxes. Livermore sought to prevent governments from learning something about the nature of the world, not about specific individuals. He wasn't trying to protect individual privacy, he was trying to limit Congress.

A 2002 Census Bureau monograph on confidentiality and privacy arrived at a similar conclusion:

> The first 50 years of census taking is most notable for the absence of concern about confidentiality. . . . A few individuals did oppose the first census in 1790, and enumerations in the early 1800s on religious grounds. They cited the Bible (II Samuel

24:1–15), where it is said that King David's taking of the census of Israel and Judah resulted in an epidemic that killed 70,000, as a reason to refuse giving any information. They also pointed out that the unwelcome results of other biblical censuses were military service and taxes. Apart from this opposition though, little if any evidence can be found that Americans were afraid of what their government would do with this personal information once they provided it to census takers.[26]

Growing Privacy Concerns

Over the decades, the Census has collected increasingly detailed information about individuals and business establishments. Eventually it became clear that many would not answer questions without an assurance that their answers would be kept confidential. Whereas the Census returns for the first two censuses were publicly posted in "two of the most important places" in each enumeration district, Congress instructed the census takers of the 1840 Census to assure the respondents that no names of individuals or companies would appear in the published statistics.[27] The legislation authorizing the 1850 Census prohibited posting returns; the 1880 Census Act required those collecting data to take an oath that they would only disclose the collected "statistics of property or business" to their supervisors and no one else; the 1890 act expanded the oath to cover "any information" at all—that is, to protect both businesses *and* individuals.

These growing concerns about privacy and confidentiality mirrored what was happening elsewhere in the nation. The Bill of Rights amended the US Constitution in December 1791 to include a specific right to privacy in the Fourth Amendment, which protects "persons, houses, papers and effects, against unreasonable searches and seizures." In the decades that followed, new technologies and business practices brought other concerns about privacy into the public's mind. Chief among these were photography (1822), credit bureaus (1841), and the low-cost paper known as newsprint (1844).

In 1890, Samuel Warren and Louis Brandeis published an influential article in the *Harvard Law Review* outlining an evolving "Right to Privacy" in US law.[28] The article identified the need to protect "persons with whose affairs the community has no legitimate concern . . . being dragged into an undesirable and undesired publicity." Unlike the laws of libel and slander, asserting that a publication is true and was made without malice should not afford publishers a defense against the charge of violating someone's privacy, the Boston lawyers argued.

A few years later, a teenager in Rochester, New York, discovered that a photograph of her taken by a local photographer had somehow ended up on a poster that was being used to sell flour; twenty-five thousand copies of the advertisement had been distributed in the United States. Abigail Roberson's guardian sued Franklin Mills, the flour producer, and the printer, Rochester Folding Box Company. She was awarded $15,000 by the Superior Court, but that award was overturned by the New York Court of Appeals, which ruled four to three that the girl had not lost anything of monetary value. The New York State Legislature responded by creating sections 50 and 51 of the New York State laws, giving New York citizens the legal right to control the commercial use of their photographs. It was the first modern privacy law, and it still exists.

In the twentieth century, the growing use of new technologies such as photography, electronic microphones, wiretaps, lie detectors, psychological testing, and increasingly large electronic data banks kept bringing privacy back into the public's concern.

The 1940 Census and Japanese Internment

Dan Bouk's book *Democracy's Data* tells the story of the 1940 Census, starting with a meeting on March 3, 1939, where business leaders from New York State met with government leaders in Washington, DC, to decide the questions that would be asked of every person and household in the United States.

These "question men," as Bouk calls them, had goals for the 1940 Census that extended far beyond the constitutional mandate of apportionment.[29] Real estate developers and retailers wanted to know where populations were concentrated and how much they were making so that they could plan new projects. Life insurance companies wanted to know the distribution of ages for setting insurance premiums. Social scientists wanted to collect information to understand mobility. The US military needed to know how many people would be available to produce armaments and fight when the United States inevitably joined the war that was unfolding in Europe and Asia.

A year later, enumerators working for the sixteenth census walked the streets of the nation's communities. Each carried population schedules that collected fifty distinct pieces of information, including each person's address, name, relationship to the head of household, sex, race, age, marital status, education, country of birth, citizenship, residence as of April 1, 1935 (to make it possible to track migration within the United States), various aspects of their employment, income, where their father and mother had been born, the language spoken at home when each person was a child, veteran status, and participation in the fledgling Social Security system. Women further were asked their number of marriages, age at first marriage, and the number of children ever born (excluding stillbirths).[30]

The Japanese Empire launched a surprise attack against the US military base at Pearl Harbor on December 7, 1941, bringing the United States into World War

II. The Census Bureau jumped into action the very next day, producing a set of tables detailing the number of Japanese citizens and people of Japanese descent living in every census tract of the United States. "The Bureau shared those figures with the military, with the FBI, and with local authorities," recounts Bouk. Census tracts are geographic areas that typically include between three and six thousand people. When the Japanese tables were finished, the Census Bureau produced "similar tabulations for Germans and Italians, even before Germany or Italy declared war on the United States."[31] The US government used those tables to plan the wartime internment of more than 125,000 individuals, most of whom were US citizens, in seventy-five concentration camps.

Although the 1941 reports included only counts, after the passage of the Second War Powers Act of 1942, the Census Bureau also released the names of Japanese Americans living in California and six other states.[32] This use of confidential data was the most egregious violation of trust and professional ethics in the history of the US Census Bureau, and it remains forever a blot on the agency's past and reputation. Yet it wasn't until the 1970s that public attention in the United States started to consider the tremendous injustice that had been wrought on US citizens of Japanese descent, and it wasn't until the passage of the Civil Liberties Act of 1988 that the survivors were granted reparations.

Figure 8 Census data powered the forced relocation and incarnation of more than 125,000 people of Japanese descent in World War II.

Microdata

The Census Bureau continued to look for more ways to make its data useful to as broad an audience as possible in the years that followed World War II. Computers made all the difference.

On March 31, 1951, the Census Bureau took delivery of the UNIVAC, the world's first commercial computer. Computers made it possible to produce more complex tabulations and create them dramatically faster than the previous system, which stored information on punched cards made from cardboard. Even more important, the diffusion of computers into the academic and business communities opened up the possibility of distributing raw data and allowing data users to perform their own tabulations.

"For the first time, the Bureau has developed for public use a sample of the individual census records of the population of the United States," reads special report 12 of the US Census of Population and Housing for 1960. "The sample is available either on magnetic computer tape or punch-cards and contains the separate records for each of approximately 180,000 persons, comprising a 0.1-percent sample of population of the United States."[33]

The sample was prepared to support social scientists, the Census Bureau said. It added, "Confidentiality of the information, as required by law, has been maintained by the omission of certain identification items."[34] The precise mechanisms were not disclosed. The cost was $1,500 (approx. $15,500 in 2023 dollars) for tape or punch cards, plus shipping. The bureau also offered to create new tabulations of previously "unpublished statistics" on request (for a fee)—a service the Census Bureau still offers to this day.

This was the first large-scale release of what is now called a public use microdata sample.

What the 2020 Census Collected

The 2020 Census collected seven pieces of information for each person in the United States:

- Name (which was not reported in any official statistics).
- Address (used to map each person to a specific record in the Master Address File).
- Age and date of birth (really the same piece of information, but both were asked).
- Sex (either male or female).
- Self-identified race. The major race categories were White; Black or African American; American Indian or Alaska Native; Asian; Native Hawaiian or Other Pacific Islander; and Some Other Race. Each of these had many subcategories. People were allowed to check as many race boxes as they wished.
- Whether each person had a Hispanic or Latino ethnicity.

- Each person's relationship to the head of household, such as being the head of household, or being that person's spouse, their child, or so on.

You can find the exact questions asked and the list of all race codes on the Census Bureau's website. The race categories were left over from the 2010 Census; the Census Bureau had planned to change the race question by adding a new category for people of Middle Eastern and North African descent, but this was not approved by the Trump administration. In 2021, the Census Bureau started asking about sexual orientation in another survey, but we don't know if it will be on the 2030 Census. (For example, in 1990 the data from same-sex couples who reported being married were edited by changing the sex of one of the individuals to be the opposite sex; in 2000 and 2010, the relationship status of the same-sex partner was changed from "married" to be an unmarried partner.)[35]

Protecting Data against Misuse

I visited the University of Pennsylvania in 2018 to give a talk about the adoption of differential privacy for the 2020 Census. After my talk, a graduate student asked me if differential privacy would prevent the US government from doing to him with his 2020 Census data what we did to residents of Japanese descent with their data from the 1940 Census—that is, would differential privacy prevent us from giving his name and address to the authorities if Congress decided to change the laws?

The student was a Muslim from Pakistan, and the then president of the United States was well-known for his anti-Islamic rhetoric. This student feared that a truthful answer on the 2020 Census might result in his being arrested at some point in the future.

"No," I told him. "Differential privacy won't protect your data from the US government. That's what laws are for."

The student didn't trust the laws. He said that laws could be changed in 2021, as they had been changed in 1942.

In fact, a technology called homomorphic encryption (HE) makes it possible to collect data with privacy protection that cannot be undone even if laws are later changed. HE makes it possible to do using math encrypted data without decrypting the data. Instead, the results of the math are themselves encrypted, although they can be decrypted later by an authorized party. With HE, it is even possible to design a system that encrypts data as each record is collected and then only allows specific analyses to be performed. The catch is that all the rules for data editing and specific allowable statistical computations need to be decided in advance and preprogrammed, which eliminates the possibility of using the microdata for many kinds of research. HE, had we decided to use it for the

2020 Census, could thus prevent the US government from ever releasing the unencrypted data while allowing us to perform a census and publish useful statistics. That would be a problem under US law, however, which calls for the release of raw data collected for each decennial census after seventy-two years to benefit academic researchers and family genealogists.

We lack technology to mathematically transport data seventy-two years into the future such that it cannot be read today. So right now, protection against misuse depends on the laws, not on the technology.

Why Is Differential Privacy Important?

In this introduction, I've tried to cover the basic ideas of why it is important to have data privacy in statistical databases, and why differential privacy is an important solution to a decades' old problem. But in doing so, I've inadvertently undersold both the increasing importance of privacy-protecting data analysis and the promise of differential privacy to help assure data privacy in the twenty-first century.

The dramatic growth of the internet and the rise of tech giants like Google, Facebook, and Amazon have demonstrated the significance and commercial value of data—especially massive amounts of confidential, highly

Figure 9 Differential privacy provides for privacy protection when data are published, but it doesn't prevent organizations from misusing data collected under a pledge of confidentiality.

detailed data about people. Likewise, local, state, and national governments worldwide are becoming increasingly adept at working with large datasets to improve service delivery, evaluate the effectiveness of existing programs, and plan new interventions. Evidence-based policymaking requires using sensitive data to make public statistics that will inform the policymaking process, as called for by the Foundations for Evidence-Based Policymaking Act of 2018.

At the same time, people are increasingly concerned about the existence of their information in these datasets. In recent years, we have seen an epidemic of targeted scams and outright theft powered by stolen data. Today's consumers want strong privacy assurances, and legislatures are responding by writing these wants into law.

The move from survey-based statistics to statistics based on administrative data makes privacy protection even more important. People filling out surveys aren't always truthful; they will (and do) skip questions and provide incorrect answers to protect their private information.

When the information comes from administrative records, such as tax records, driver's license data, or naturalization records, people don't have a choice: their personal information is used without their consent—and frequently without their knowledge. As a result, it is especially crucial in these situations that the statistical products have appropriate privacy protections.

Figure 10 A primary use of differential privacy in the US government will be making evidence in line with the goals of the Foundations for Evidence-Based Policymaking Act of 2018.

Because of its mathematical underpinnings, differential privacy is increasingly being promoted as a way to use sensitive records for statistical and research purposes without obtaining the consent of the data subjects. The theory is that there is no need to request consent if the data subject's privacy is being properly protected.

As we will see, there are important mathematical limits to this whole idea of anonymization. Nevertheless, to the limits of what's mathematically possible, differential privacy's approach for privacy accounting is the only workable technique that's been developed to date.

1

CONCEPTS AND THEORIES

The last chapter provided a broad overview of differential privacy with a simple example of database reconstruction and the historical context of the US Census. This chapter will drill down on those concepts. It will involve a little math, but not much.

DP in a Nutshell

To start with, here is my take on the core concept of differential privacy:

> *Adding noise* to the result of every database query is the only way to provide for *composable confidentiality protection* in a *statistical database*. Differential privacy is a mathematical *framework* based on a definition

of *privacy loss* that has *formal guarantees*. In realizing those guarantees, differential privacy makes it possible to control the *trade-off* between privacy loss and *statistical accuracy*.

The remainder of this section will go through each of these terms and use them to explain the basics of how DP works.

Adding Noise

The idea of noise as a quantifiable thing, rather than just an annoyance that makes it hard to understand spoken words, comes from the 1920s, when engineers at the American Telephone and Telegraph Company (now called AT&T) started researching the signal and noise of early transatlantic radio links.[1] Noise was a problem for these early engineers, limiting the usefulness of their new communications systems. But try as they might, the engineers discovered that they could not conquer noise; they had to live with it. As part of this effort, they developed statistical models that described noise as an ever-changing, random variable.

There are many kinds of random variables. Take a coin, toss it in the air, and catch it in your hand. Now open your hand; on top is either the coin's face (the *head*) or its back (the *tail*). Repeat eight times and you have eight independent random draws: H H T H T H T T.

Analyzing sequences like this is the basis of probability theory, a branch of mathematics and important part of statistics. Here each coin toss is *independent*, meaning that we cannot do a better job predicting the next coin toss by knowing the previous one. Ideally a coin toss should also have a *uniform distribution* because each coin toss has the same chance of the two choices. This is sometimes called a *fair coin*. Above we flipped the coin eight times, and got four heads and four tails. If we try again, we might get H T H T H H T H—five heads and three tails. We call this set of eight coin tosses a *trial*.

Being fair doesn't mean that we always get roughly the same number of heads and tails. If we repeat this experiment a thousand times, sometimes we might not get any heads at all, and at other times we might get eight heads. If we track the total number of heads for each trial, most likely we'll find that the trials with four heads were the most common. Next will be the number of trials with three or five heads, which should have similar counts. After that we'll have trials with two or six heads, then with one or seven heads. Least common will be the trials with zero or eight heads. If we graph the results, it will look like graphs of a *bell curve* that you might see in a statistics textbook.

The bell curve graph is also called a *normal distribution* or *Gaussian distribution*, after Carl Friedrich Gauss, who discovered when he was eighteen years old that the distribution matched errors in astronomical measurements.

Communications engineers realized that they could use random values drawn from a Gaussian distribution to model the noise that they were hearing on their links; they started calling this *noise*. By the 1940s, statisticians were using the term to describe a source of random values even where no telecommunications equipment was involved. Today we say that DP provides privacy by adding *statistical noise*.

While many different noise distributions can be used in DP, most textbooks use the *Laplace distribution* because it only uses the single privacy parameter ε, and because the resulting privacy guarantees and composition rules are easiest to understand. In the future most deployed systems are likely one of the DP variants, such as zero concentrated differential privacy (zCDP) because these variants tend to produce more accurate results with similar amounts of privacy loss.

De-Identification Doesn't Compose

On May 18, 1996, Massachusetts governor William Weld was at the Bentley College graduation to give the commencement address. He got up, started walking to the podium, and collapsed. The governor was then rushed by ambulance to the Deaconess Waltham Hospital, where he was diagnosed with influenza, given a stress test, and discharged the next day. He made a speedy recovery.[2] Governor Weld's tests and procedures were paid for by the

Massachusetts State Employee Group Insurance Commission (GIC) insurance plan.

Health care costs have been steadily rising for decades. One of the big hopes for evidence-based policymaking is that researchers examining health care data will be able to use data to learn which procedures are cost-effective and which are a waste of money, if not actually counterproductive. To do this, researchers need detailed health care data. Although it is possible to get these data one patient at a time by asking patients to participate in research studies, many will simply say "no" because they are concerned about their privacy. So instead, some health care providers, including the GIC, have made de-identified insurance reimbursement records available for research.

The idea behind de-identification is to take the dataset of sensitive records (in this case, health insurance claims) and remove the *identifying* data, such as each person's name and address. Sometimes the identifiers are replaced with a code number, called a *pseudonym*, so that different records belonging to the same person can still be linked together.

The GIC's de-identified records still included information that would be useful for medical research, such as sex, age, and zip code. Sex and age are obviously important for understanding health. But instead of just the patient's year birth, the GIC "de-identified records" actually included the their full date of birth, which is vital for linking together records from different providers. The records also included the patient's full five-digit zip code, allowing researchers to account for environmental factors—living near a coal-burning power plant might increase your chances of asthma, for example—and make it easy to adjust for social economic status without having to obtain information about each person's income or wealth. The five-digit zip helps with data linkage as well.

De-identification is a common privacy building block, but its effectiveness depends on things outside the control of the organization releasing the data. In the case of the GIC dataset, an MIT graduate student named Latanya Sweeney realized she could pick out the records belonging to Governor Weld if she knew his sex, date of birth, and zip code. His sex was obviously male, it was commonly reported that the governor lived in Cambridge, and she could get his date of birth from the Cambridge voter registration dataset—which the city of Cambridge sold for $20.

Sweeney graduated and embarked on an academic career devoted to protecting privacy. Her reidentification of Weld's medical records in the GIC dataset was one of the factors leading to the adoption of the HIPAA Safe Harbor Privacy Rule, which forbids releasing de-identified datasets with such detailed information.[3] Other than the reidentification of Governor Weld's medical records, Sweeney is best known for *k-anonymity*, her approach for mathematically formalizing de-identification. Today *k*-anonymity is widely used and seen as a competitor to differential privacy. But *k*-anonymity lacks a key property of differential privacy, and as a result, does not offer the same strong privacy protections.

k-Anonymity Doesn't Compose

k-anonymity is designed to work with datasets of tabular data, where each row represents a different person and each column represents a different piece of information about that person. Each row must have the same number of columns. To use *k*-anonymity, the analyst classifies each column as either an *identifier*, *quasi-identifier*, or *nonidentifying*. Identifiers, like Weld's name and street address, are removed from the dataset. Nonidentifying information, like the medical tests that were run and their results, are left in the dataset. This leaves the columns that can identify a person when used in combination. Sweeney's idea was that for any combination of quasi-identifiers, there should be at least k people who shared that exact same combination.

Applying Sweeney's *k*-anonymity to a dataset creates an *anonymity set* of k people for any possible reidentification attempt. Here k is the measure of privacy protection. Larger values of k produce more uncertainty. If $k = 2$, then for any record in the dataset there is a 50 percent chance of linking the records using some external dataset. If $k = 100$, then there is only a 1 percent chance.

Returning to the GIC dataset, if nobody else in Cambridge had the same birthday as Weld (July 31, 1945) and lived in his ZIP Code (02138), protecting this dataset with *k*-anonymity would require generalizing Weld's birthday to the year and month (July 1945) and perhaps generalizing the ZIP Code to the first three digits (021-). If *k* people still did not match, perhaps the birthday would be generalized to just the year (1945) or a range of years (1940–1945). Eventually there would be *k* people with the same combination of quasi-identifiers as Governor Weld. The theory is that if *k*-anonymity had been applied in advance to the GIC data, it would have been harder (if not impossible) for Sweeney to match Weld's medical data with his name on the voter rolls using the age, sex, and ZIP Code quasi-identifiers.

One problem with *k*-anonymity is that it implicitly assumes that there exists data in the dataset that are not identifying. In practice, what makes data identifying is that they can be linked with data in another dataset. That's the mosaic effect! And there's no way for the person performing the de-identification to know if another such dataset exists or not.

Consider that on May 19, 1996, Weld was probably the only person at the Deaconess Waltham Hospital to have two attending cardiologists and receive an electrocardiogram, blood enzyme test, and echocardiogram—facts that were reported the next day in the *Boston Globe*.[4] Normally

Figure 11 By cross-referencing databases, many kinds of de-identified data can be reidentified and confidential data potentially revealed.

such information wouldn't be identifying, but for Weld it was, because of the existence of that newspaper article.

This is a problem for more people than the governor of Massachusetts. In 2011, Sweeney discovered that supposedly anonymized patient-level hospital health records distributed by the state of Washington could be reidentified using newspaper accounts of accidents. "A single searchable news repository studied uniquely and exactly matched medical records in the state database for 35 of the 81 news stories found in 2011 (or 43 percent), thereby putting names to patient records," she wrote in a paper describing that project.[5]

k-anonymity is not *composable*.

Differential Privacy Is Composable

Differential privacy is composable. If an organization uses differential privacy to produce five different reports from a confidential dataset, the organization can mathematically compute the maximum possible privacy loss that might result if an attacker manages to combine all of those reports in an effort to learn the confidential data.

k-anonymity is not composable; it might be possible for an attacker to combine two extracts of a confidential database created with $k = 10$ and produce a new dataset with $k = 1$—that is, a dataset that potentially identifies individuals. There's no way to tell in advance; it depends on the specific datasets and how k-anonymity was applied.

The inability of k-anonymity to compose was at the root of a 2014 lawsuit against the US Department of Veterans Affairs in which the government blocked the release of a de-identified dataset because it could be combined with other de-identified datasets to undo the privacy mechanism.[6]

Even worse, a single $k = 10$ dataset might still make it possible to identify individuals. That's because k-anonymity doesn't add noise to the "nonidentifying" values. But there might be a second dataset—for example, a newspaper article—that prints enough information about a person that it's possible to identify their record using the supposedly nonidentifying values. Unlike records processed with DP, with k-anonymity there is no deniability for the individual attributes in the dataset once an identification is made because noise is not added.

Confidentiality Protection

DP is part of a long tradition of mathematical techniques designed to protect confidential data about individuals while allowing the data to be used in aggregate.

The most straightforward way to protect confidentiality is by not collecting confidential data in the first place, throwing the data away when it is no longer needed, or not making statistics from confidential data. These are all examples of *data minimization*. While minimization is an important way to protect privacy, it's not the focus of this book.

Originally statisticians thought that they could protect confidentiality by simply not including the names of people and establishments in their statistical reports. This is the essence of de-identification: simply remove the identifying information. Over time this simple approach caused problems as reports became increasingly detailed.

For instance, reporting how many people are employed in each state making steel is not much of a problem if there are more than a handful of steel mills. But what if there is only one? In that case, a state-by-state table reveals how many people were employed by a specific company. If there are two companies, they are probably competitors; a report of their combined employment lets one determine the other's confidential data.

In 1910, President William Howard Taft issued a proclamation noting that April 15 was Census Day and reminding the population that "it is the duty of every person to answer all questions on the census schedules."

The proclamation continued,

> The sole purpose of the census is to secure general statistical information regarding the population and resources of the country. . . . The census has nothing to do with taxation, with army or jury service, with the compulsion of school attendance, with the regulation of immigration, or with the enforcement of any national, state, or local law or ordinance, nor

can any person be harmed in any way by furnishing the information required. There need be no fear that any disclosure will be made regarding any individual person or his affairs.[7]

As the century proceeded, statisticians at the Census Bureau had an increasingly difficult time making good on Taft's promise. For the 1930 Census, the US Census Bureau stopped publishing so-called small-area data because it had no technology for preventing the indirect disclosure of information from those tables.[8] Instead, some statistics were printed with blanks in particular cells—an approach called *suppression*.

A common suppression approach is the so-called *rule of three*: results from three or fewer respondents are suppressed. This rule was used as recently as 2005 by the agricultural estimates program at the National Agricultural Statistics Service.[9] The service also used a technique called the *(n,k) dominance rule*, in which a cell is suppressed if a combination of n respondents accounts for more than *k percent* of the cell's value. Here the values n and k are tunable parameters. For example, if $n = 1$ and $k = 0.6$, then the cell would be suppressed if any firm contributed to more than 60 percent of the answer. This is called *primary suppression*.

Primary suppression is rarely sufficient to protect confidentiality. If a cell in a table with row and column totals is suppressed, then other cells need to be suppressed so that the missing data cannot be readily recalculated. Generally the row and column totals can't be suppressed because policymakers as well as the public want to see the totals. Instead, other cells need to be suppressed—ideally ones that have small and inconsequential values. This is called *secondary suppression.*

Dalenius's 1977 article codified the methods and objectives of suppression and other approaches to statistical disclosure limitation.[10] Recall that his goal was that nothing should be learnable about an individual or establishment from a published statistical table without using the table. Dwork's 2006 paper "Differential Privacy" showed that these desiderata are impossible because of auxiliary information—that is, the mosaic effect.[11]

But even without auxiliary information, the task of suppression becomes increasingly more complex as an organization wants to make more use of its confidential data. If a suppressed value appears in two tables, then each of those tables requires their own suppression patterns. The two tables might interact with each other, however, requiring even more cells to be suppressed. What's worse, if one value of the suppressed cells can be learned from another data source, then it might be possible for someone to undo all the confidentiality protections. That's the mosaic effect again.

With the birth of the commercial internet in the 1990s, it suddenly became much easier to distribute data.

Simultaneously, there was increased pressure on governments and nonprofit organizations to make data available. Engineers and information technology professionals who had little knowledge of statistical disclosure avoidance had large datasets that they wanted to share: some thought that all that they needed to do to protect privacy was to strip off the names. They called this process *anonymization* because in their minds, they were removing the identifiers from the data and making them anonymous. This is precisely the problem that the Massachusetts GIC encountered. It's also why I avoid the term *anonymization* and speak instead of *de-identifying*, which literally describes the process that the data experience rather than the hoped-for end result.

Statistical Database

A database is a structured collection of data that is organized for efficient data processing. A database need not be electronic or accessed with a computer; an ordered library of books is a database. Likewise, a single printed volume containing words in the English language and their meanings is a database of word definitions; such databases are commonly called *dictionaries*.

A *statistical database* is a special kind of database that is organized for creating statistics. A stack of punch cards from the 1890 Census would be such a database, if any such cards still existed today. (The schedules from the

1890 Census were punched onto sixty million punch cards, but most of the schedules and cards were destroyed in two fires—in 1896 and 1921. Finally, in 1933, Congress approved a request from the Bureau of the Census to destroy what remained.)[12]

Statistical databases are not limited to the US census or even government records. During the COVID-19 pandemic, Google took its vast collection of personal data from the hundreds of millions of people all over the world who were running the Google Maps application on their cell phones and created a single statistical database containing each person's day-to-day movements. Google then used this statistical database to create a series of "COVID-19 Community Mobility Reports" to help researchers determine the actual impact of various stay-at-home orders, closings of schools and workplaces, and other "lockdown" policies.[13] Instead of asking users to opt-in, Google used differential privacy to aggregate and protect data confidentiality.[14]

Other examples of statistical databases include health care records, financial information, consumer purchases, and education records. In the 1960s, there was a serious proposal to build a single "national data center" that would contain tax, education, military, employment, and other records for all people living in the United States to aid in statistics making for government, academic researchers, and even business.[15] After much discussion and several hearings in Congress, the project was abandoned due to privacy concerns.

Differential Privacy Is a Framework

Differential privacy is a mathematical approach for statistical disclosure control, but it is not a single mathematical algorithm or process. Instead, DP is a framework based on a formal definition for privacy loss. It's this formal definition that is the breakthrough that Smith presented at the 2006 conference.[16] The formal definition appears in figure 12. Since this is math, different publications can present this same equation with different variables and sometimes rearrange them, but the equation always means the same thing.

Let's take this equation apart and figure out what it means:

- In this version of the equation, D_1 represents a dataset that has your data and D_2 represents a dataset that doesn't have your data. Differential privacy calls these *neighboring datasets* or *neighboring databases*. To use the example from the previous chapter, D_1 might be the dataset with ten grades in the class, including yours, and D_2 might be the grades of the nine other students (all of whom got 100), absent your grade (which was an 80). In other versions of the formula, the variables D and D' are sometimes used to denote the two datasets.

Figure 12 Differential privacy is a formal definition of privacy loss.

The mathematical definition of differential privacy is based on DP's key insight, as mentioned earlier: an analysis performed on a confidential dataset that doesn't contain your data can't impact your privacy. Thus a privacy-preserving computation based on a dataset is one that produces the same result whether your data is present or not. (See the section "Differential Privacy's Key Insight" in the introduction.)

The second part of this key insight is that D_1 and D_2 don't represent a specific dataset that's being protected; they represent every possible dataset to which a specific differentially private analysis might be applied. So when we analyze the grades in the class, we don't just protect the student with the 80; we also protect the nine students who got 100. But our analytic approach would also work if the students got other grades. In fact, it would work with *any possible distribution* of grades.

- The function $\mathcal{A}()$ is the differentially private query function: it accesses the dataset and returns an answer.

In the lexicon of differential privacy, any function that implements a differentially private query is called a *mechanism*. The stylized $\mathcal{A}()$ in the definition means *all such mechanisms*. The DP formula is sometimes written with an **M()** instead of an $\mathcal{A}()$.

The definition of differential privacy doesn't tell us how to create a mechanism. Rather, it gives mathematicians a way to determine if a proposed mechanism is differentially private or not: if the mechanism can be plugged into the definition and then a mathematician can prove that the inequality in the cartoon still holds true, the proposed mechanism is differentially private.

This is one of the advantages of having a formal privacy definition. We don't need to try it out on real data and see if it protects privacy. All we need to do is to see if the mechanism satisfies the inequality. This means that scientists throughout the world can develop their own customized approaches for protecting statistical data. They can then prove that their approach works by showing it follows the rules of DP. With that out of the way, we can evaluate the usability of the resulting data, and plot the trade-off between privacy protection and data utility.

- The ***S*** represents a set of possible answers that the query function can give for any given dataset. When mathematicians check a proposed mechanism against the definition, they don't try it out for a specific set of data; instead, they prove it for all possible datasets.

- $\boldsymbol{\varepsilon}$ (epsilon) is the privacy loss parameter. The value can be anything between 0 and ∞ (infinity). As with *S*,

Scientists throughout the world can develop their own customized approaches for protecting statistical data. They can then prove that their approach works by showing it follows the rules of DP.

we don't check the equation for a specific value of ε. By design, DP mechanisms can be tuned for any value of maximum privacy loss.

- ***e*** is the mathematical constant known as Euler's number. Like π (pi), e is an irrational number; the first eight digits of e are 2.7182818; it continues forever without repeating.

- **Pr[. . .]** is the probability function. Pr[x] is the probability that x is true, x is going to happen, or x has happened. In some probability textbooks, the probability function is written *P(x)*, but in the differential privacy literature it tends to be written Pr[. . .].

The DP definition says that we should consider the probability that the function $\mathcal{A}()$ applied to dataset D_1 will produce roughly the same result as same function $\mathcal{A}()$ applied to dataset D_2. If the ratio of those two probabilities is less than e^ε, then $\mathcal{A}()$ implements a differentially private analysis. The equation defines "roughly the same" to mean that the ratio of the two probabilities will be less than a small number—at least when e^ε is a small number (which happens when ε is significantly less than one. The symbol e^ε is sometimes written as *exp(ε)* because that's somewhat easier to read—and typeset.)

If you find the idea of taking the ratio of two probability functions confusing, it may be helpful to think about the statistics that you can get from rolling a pair of six-sided dice. The first die can have any of six faces pointing face up (⚀⚁⚂⚃⚄⚅) and so can the second. This means that there are 36 combinations that a pair of dice can role. The interesting probabilities come when you add the number of dots on each die together: because the dice can only sum to the numbers 2 through 12, there is an unequal distribution of possible outcomes.

For example, there are six ways for two dice to role a 7 (⚀⚅, ⚁⚄, ⚂⚃, ⚃⚂, ⚄⚁, and ⚅⚀), so the probability of rolling a 7 is ⅙, making Pr[*rolling* 7] = ⅙. There is only one way of having the sum add to the number 2 (⚀⚀), which makes the odds 1/36. The same is true with the odds of rolling 12 (⚅⚅). Thus Pr[*rolling* 2]/Pr[*rolling* 12] = 1.0, but Pr[*rolling* 2]/Pr[*rolling* 7] ≅ 0.166.

Now let's go back to the parameter ε (epsilon) and the constant e. If $\varepsilon = 0$, then $e^{\varepsilon} = 1$. This means that the probability that $\mathcal{A}()$ produces a value in set S is the same for both D_1 and D_2.

If the probability of outputs from $\mathcal{A}(D_1)$ and $\mathcal{A}(D_2)$ are the same, then query function $\mathcal{A}()$ can't distinguish between the two datasets. But the datasets are in fact different; by definition, they are neighboring datasets, having different data for *one person*. $\mathcal{A}()$ ignores that difference when $\varepsilon = 0$. And since that one person can be any possible person in the universe, $\mathcal{A}()$ must be ignoring the personal data for all possible people. This is why there is

Figure 13 Epsilon is pure differential privacy's single tunable parameter. All hail epsilon!

no privacy loss when $\varepsilon = 0$; the query function $\mathcal{A}()$ ignores differences between people in the input dataset.

As ε gets larger and larger—as it marches toward infinity—the ratio of the two probabilities can increase as well. This means that there is an increasing chance that $\mathcal{A}(D_1)$ and $\mathcal{A}(D_2)$ produce different results: one of them produces a result that is in S, while the other does not. This means that there is an increasing chance that the outputs of $\mathcal{A}(D_1)$ and $\mathcal{A}(D_2)$ can be distinguished. This increasing risk of distinguishability is DP's concept of privacy loss.

Sensitivity

To use the DP mechanism, you must determine the sensitivity of a particular query. *Sensitivity* is the range of possible influences that one person's data can have on the final measurement. It depends on the design of the data pipeline rather than the confidential data that are being analyzed. In our classroom example, the average for a class with ten students has a sensitivity of ten because any one student's quiz grade can swing the class average by as much as 10 points.

The amount of noise added by DP is proportional to the sensitivity of a computation, so in practice a lower sensitivity means a lower PLB for the same level of accuracy. For instance, if there are twenty students in the class, the sensitivity of the class average calculation drops to 5 points, and if there are a hundred students, it drops to 1. Thus for the same amount of noise, the accuracy of the class average computation increases as the number of people being analyzed increases given that the sensitivity goes down.

Mathematically, we use the symbol Δf for sensitivity, where Δ is the Greek letter delta, which typically means "change" in mathematics, and f is the query function.

Computing the sensitivity of a query is easy in some cases. For the 2020 Census, most queries were *counting queries*, used to produce tables that described how many people met a certain condition—such as how many people eighteen or over lived on a particular census block, how many of those people had indicated they were White on their census form, how many said they were Black or African American, and so on. Each of these queries had a sensitivity of 1.

The sensitivity of a query can be hard to compute in other cases. An economist that I worked with wanted to compute the sensitivity for the coefficients of a linear regression. This presented a problem because a line might be horizontal or vertical, making the sensitivity of the coefficients effectively infinite. That doesn't mean that it's impossible to make a regression that's formally private; it just means that naive approaches don't work. Indeed, designing high-quality differentially private regression mechanisms is an area of active research.[17]

Once we have computed the sensitivity, we need to pick a value for ε, the privacy loss parameter. It turns out picking a value of ε is hard. In the early days of differential privacy, many researchers thought that a value of 1 represented a good trade-off between accuracy and data privacy. Today's thinking on the topic is more sophisticated: we know that every possible value of epsilon represents a compromise between two competing goals, and setting the value of epsilon is a decision for people implementing policy, not for backroom technologists.

The Laplace Distribution

In addition to introducing a mathematical definition for privacy loss, the 2006 paper presented a mechanism that

We know that every possible value of epsilon represents a compromise between two competing goals, and setting the value of epsilon is a decision for people implementing policy, not for backroom technologists.

satisfies the definition. That mechanism is called the *Laplace mechanism* because it is based on noise drawn from the Laplace distribution.

In statistics, the word *distribution* is sometimes used to describe a random sequence of numbers. (I sneaked this bit of math terminology into a sentence in the previous section.) The simplest distribution is the *uniform distribution*, which is a distribution for which each number is equally probable. If you roll a single die, there is a ⅙ probability that the next roll will be a ⚀, ⚁, ⚂, ⚃, ⚄, or ⚅. If you roll two dice and add their values together, repeat this experiment a few times, and graph the number of times you get each value, you produce a distribution that looks like a triangle (and that is called a *triangular distribution*). As you increase the number of dice, the distribution looks less like a triangle and more like the famous "bell curve," properly known as a Gaussian distribution. The distribution looks really bell-like if you make many rolls of a hundred dice.

Statisticians like the Gaussian distribution because it is mathematically easy to work with and seems to do a good job approximating some things in the physical world, such as the errors that astronomers make when measuring the location of the stars or planets—what Gauss was doing back in 1795.

The Laplace mechanism, to repeat, is based on the Laplace distribution. Whereas the Gaussian distribution has a rounded top, the Laplace distribution comes to a sharp point. Whereas the Gaussian distribution is tight and well contained, the Laplace distribution has broad "tails" on either side of its central value.

As an experiment, I asked my computer to generate five numbers from each distribution using its default parameters. The results, in table 2, shows that the values from the Laplace distribution tend to be more extreme than those from the Gaussian distribution.

With either distribution we could take thousands of numbers, average them all together, and end up with a result that's close to zero. If we averaged a million numbers, we would be even closer to zero. Yet the Gaussian distribution will produce an average closer to zero because the tails are smaller.

The Laplace Mechanism

Once we know the desired PLB (ε) for a query and that query's sensitivity (Δf), the Laplace mechanism scales the amount of noise to add to each query by the value $\Delta f / \varepsilon$:

Noise to add $= \mathrm{Lap}(\Delta f / \varepsilon)$

Thus for any query function f applied to dataset D, we can protect the results with differential privacy by adding noise:

Protected f(D) $= f(\mathrm{D}) + \mathrm{Lap}(\Delta f / \varepsilon)$

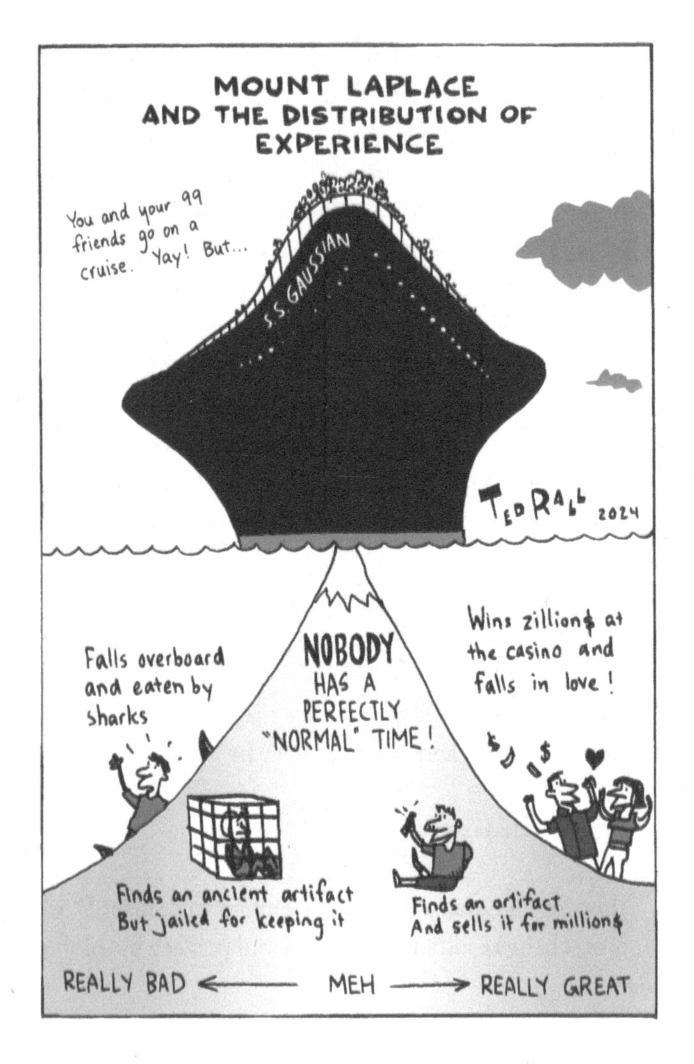

Figure 14 The Laplace distribution has many more low-probability events than the Gaussian distribution.

Table 2 Five Draws from the Gaussian and Laplace Distributions

Gaussian distribution	Laplace distribution
-1.256	0.145
-0.106	0.236
0.929	-1.767
-0.201	-2.840
-0.520	0.908

This means that a counting query—for example, *How many people are on this block?*—becomes differentially private by adding Laplace noise scaled by $1/\varepsilon$. Notice that as ε approaches zero, the value of $1/\varepsilon$ approaches infinity, as does the amount of noise that is added. The noise completely dominates whatever value $f(D)$ returns. And as ε gets larger from zero, the value of $1/\varepsilon$ gets smaller, to the point that no detectable noise is added. With no noise, the Laplace mechanism simply releases the unprotected value of the query function.

This is why it is never enough for an organization to say that it is protecting data with differential privacy; it's always necessary for the organization to provide the value of epsilon and the mechanism that it's using. This isn't an ideal outcome, but it's better than the statistical approaches for protecting confidentiality that came before differential privacy.

Composition Rules: Parallel Composition

Many queries that data scientists perform on confidential data follow a straightforward pattern that I call *filter*, *compute*, and *output*:

- **Filter:** First, the query selects which rows in the dataset to process by filtering the data based on a selection criterion. In the Census, for example, many of the statistics are reported block by block, so the first thing we do is filter the data based on a specific block. A query for the number of people over eighteen on that block additionally filters by age.

- **Compute:** Next the query takes the filtered data and uses them to perform a computation. For instance, it might count the number of people in the filtered set.

- **Output:** Finally the query outputs the results. This output can be shared with the public, shared internally within an organization, or used as a building block for other queries.

Differential privacy's composition rules describe how the total PLB changes when multiple queries run against the same confidential dataset. These rules are largely based on how the filter steps are composed into the data pipeline.

When designing electronic circuits, components like batteries, lights, and switches can be wired in *parallel* or *series*. The same is true of flows in data pipelines.

Note that in many cases, only a single analyst sees the output, after which the output data are deleted. This happens when analysts are exploring the data. Differential privacy doesn't handle data exploration gracefully; I discuss some of the issues in the next chapter under "Data Exploration."

Parallel composition happens when several queries operate on nonintersecting samples of the confidential data. For example, a query that determines the average age of all men in the sample composed with a second query that determines the average age of all women is an illustration of parallel composition. If the dataset is used for two queries that have the same value of ε, then the total privacy loss is also ε. If the two parallel queries have different values of privacy loss, ε_1 and ε_2, the total privacy loss is the larger of the two—$\max(\varepsilon_1, \varepsilon_2)$.

Here are some examples of parallel composition when taken together:

Parallel composition example #1, query #1

- ✔ Queries used to create a table that shows the population for every block in a city (assuming that the blocks do not overlap).

Parallel composition example #1, query #2

- ✔ Queries for people in the sample divided into age ranges that don't overlap, such as 0–5, 6–17, 18–21, 22–64, and 65–115. As this example demonstrates, the filtering need not produce samples of the same size or that cover the same range, as long as the samples do not overlap.

And here are some instances for queries that are not parallel composition when taken together:

Not parallel composition example #2, queries #1–#2

- ✗ A query for men over 18 and a query for men over 18 with a college education.
- ✗ A query for men over 18 and a query for people over 18 with a college education.

Parallel composition allows an analyst to create a data pipeline that produces many answers with only a single drawdown on the overall PLB, so analysts building a data pipeline try to use parallel composition as much as possible. For example, a data analyst could rewrite the previous queries that are not parallel composition into four queries that have parallel composition, saving privacy budget in the process. The rewritten queries might be:

Parallel composition example #3, queries #1–#4

- ✔ Number of men over 18 with a college education.
- ✔ Number of women over 18 with a college education.

- ✔ Number of men over 18 who do not have a college education.
- ✔ Number of women over 18 who do not have a college education.

Because an estimated 1.7 percent of the population has an intersex trait, and "approximately 0.5 percent of people have clinically identifiable sexual or reproductive variations," two additional queries could be added as well while maintaining parallel composition:[18]

Parallel composition example #3 continued, queries #5–#6, option #1

- ✔ Number of people identifying as neither male nor female over 18 with a college education.
- ✔ Number of people identifying as neither male nor female over 18 without a college education.

Because the number of intersex people is tiny compared to the overall population, it might make more sense to combine these two queries into a single one:

Parallel composition example #3 continued, query #5, option #2

- ✔ Number of people identifying as neither male nor female over 18.

Combining the two intersex queries into a single query produces a protected answer that is more accurate because the Laplace noise is only added once. This is a decision, however, that the analyst needs to make during the design of the data pipeline rather than after seeing the confidential data. That's because with differential privacy it is not permissible to review the results using live data and use those results to produce a better pipeline. While such iterative tuning is common traditional statistical analysis (at least for analysts not versed in formal privacy), it produces undesirable feedback. In traditional data modeling, this feedback is sometimes called *overfitting*, resulting in potentially amazing model results for a specific dataset, but poor results for others. When using differential privacy, the feedback produces *unaccounted privacy loss* or *privacy leakage*. This is because the repeated test queries have improved the result in a way that is apparent to an outside observer. This is an illustration of what's known as the *Fienberg problem*.

Composition Rules: Sequential Composition

Multiple queries that do not enjoy parallel compositions are said to use *sequential composition*.

If you wire three similar light bulbs in a house connected to the same switch, you will typically wire the bulbs in parallel: a wire goes from the switch to each bulb, and then from the bulb to the neutral. When you then flip on the switch, the full voltage goes to each bulb, and each bulb will be equally bright. But if you mistakenly wire the bulbs in series, each bulb will be dim, if they light up at all. That's because each bulb is now only getting only a third the voltage that it needs.

Differential privacy's sequential composition works in a similar manner. If you have two queries that do not compose in parallel, then you calculate the total possible privacy loss of the two queries simply by adding together the epsilon values for each. "The epsilons add up," is what many people say.

Relaxed Differential Privacy

Until now we have been discussing a particular variant of differential privacy that is formally called ε-differential privacy (also known as "pure" differential privacy). In practice, ε-DP is rarely used in production systems because the long "tails" of the Laplace noise distribution produce significant errors for even modest amounts of privacy loss.

Production systems more commonly use variants of differential privacy that allow for better guarantees on data quality. These variants are called *relaxations* of differential privacy because their privacy guarantees are not as strict. The relaxations make it possible to produce results that are more accurate for the same amount of privacy loss under most practical circumstances. But there is no free lunch in data privacy: the relaxations come with the cost of a small amount of unaccounted-for increased privacy risk. Typically, that risk is quite small and not very relevant.

Approximate differential privacy was the first example of a DP relaxation. Also known as (ε,δ)-differential privacy, this relaxation is typically implemented using the so-called *Gaussian mechanism*, which is based on Gaussian noise.

With (ε,δ)-differential privacy, ε is still a measure of privacy loss, and higher values of ε result in less privacy protection and more accurate answers. In fact, the answers are generally more accurate than with ε-differential privacy for the same value of ε. Yet with each application of an approximate differential privacy mechanism, there is a small chance that the privacy protection mechanism will not fully protect the confidential data. This probability is represented by the Greek symbol δ (lowercase delta), another tunable parameter. Although δ can range from 0 to 1.0, typically we use tiny values—between 10^{-10} and 10^{-6}. The first is one chance out of ten billion, roughly the chance that a roulette wheel will spin "000" six times in a row. The second is one chance in a million, which is sixty-five times less likely than the chance of being struck by lightning.

Concentrated differential privacy is another relaxation of differential privacy that similarly allows for more control over the distribution of error and privacy loss, especially when there are multiple queries. Concentrated differential privacy was invented by Dwork and Guy N. Rothblum in 2016.[19] Later that year, Mark Bun and Thomas Steinke developed a variant they called zero-concentrated differential privacy (zCDP), which is better suited to many uses.[20] These variants are most often used with data pipelines that involve many individual queries that do not enjoy parallel composition, such as iterative machine learning algorithms. Google's differentially private version of its popular TensorFlow machine learning library uses zCDP.

For the 2020 Census, our first plan was to use pure differential privacy, but we switched to using zCDP several years into the project because it produced dramatically more accurate results with less overall privacy loss. The upgrade was relatively simple: the changes were largely confined to the random number generator, the privacy accounting, and new tools that were developed to communicate the new privacy loss metric to data users and the general public. That last part happened after I left.

Membership Inference

Although today privacy professionals are mostly concerned with the risk of reidentification (sometimes erroneously called de-anonymization) in public data, privacy researchers tend to be more focused on the risk of *membership inference*, which is easier to express mathematically. It can also be used as a stepping stone for reconstruction and reidentification.

Membership inference means that a data hacker can assess that a specific individual's data is part of a dataset. In the case of machine learning (discussed above), it means that the attacker can infer that the individual's data was used in training.

Differential privacy protects against membership inference attacks. This isn't surprising given that the definition of differential privacy is based on the indistinguishability of a dataset that contains an arbitrary individual and one that does not.

Points of Confusion and Clarification

This section presents some of the more confusing aspects of differential privacy that have caused problems for both privacy researchers and data scientists who have tried to deploy it in practice.

Inference about Groups Do Not Violate an Individual's Privacy

One of the things that makes population statistics possible is the fact that groups of people share common characteristics. As a result, it's possible to make predictions about individuals by merely sampling the population. For example, if you randomly sample ten thousand people in the United States and discover that roughly nine thousand of them are right-handed, then there is a good chance that any person you randomly pick in the United States will be right-handed too. If you call me up and say, "Simson, I think that you're right-handed," the fact that you can make this inference about me isn't a violation of my privacy; it's simply an application of statistics.

Discovering that 90 percent of the people in the United States are right-handed doesn't tell you if I am in fact right- or left-handed. If I considered this detail confidential, if you found out, and if you then told the world, that would be an actual privacy violation. (Using Solove's taxonomy, shown in table 1, it would be privacy harm C1, breach of confidentiality.)

Differential privacy doesn't prevent these kinds of predictions. To the contrary, what DP does is allow these predictions to be made while making it harder to know if a specific person that you know was included in the study is right- or left-handed. DP also makes it harder to know if a given person was included in the study or not (see "Membership Inference" above).

Consider a series of studies in the 1990s that discovered Ashkenazi Jewish people are significantly more likely than the general population to have mutations that increase their chances of developing cancer. One study, based on anonymized blood samples from 858 individuals, found a threefold increase in the number of Ashkenazi Jewish people carrying a cancer-causing mutation in the

BRCA1 gene when their genetics were compared with the general population. The mutation also increases the risk of getting breast cancer by a factor of 5, from 4 in 25 (16 percent) to roughly 4 in 5 (70–87 percent).[21] Another study found that approximately 6 percent of Ashkenazi Jews have a mutation increasing the risk of colon cancer from 9 to 15 percent, to 18 to 30 percent. A third study found that Ashkenazi Jews are at higher risk for schizophrenia and bipolar disorder.

Differential privacy didn't exist when these studies were conducted, but let's pretend that it did, and that it was used to protect the publication of these results with an $\varepsilon = 1.0$. With that level of privacy protection, it would be difficult to know if any given person of Ashkenazi Jewish descent living in the United States had been involved in the study or not. There were roughly 5 million Ashkenazi Jewish people in the United States at the time of the studies—2 percent of the US population—but only 858 people were study participants. With an $\varepsilon = 1.0$, we might be able to reduce the universe of possible study participants from 5 million to 1.8 million. That's because $e^{(\varepsilon=1)} \cong 2.71$. This means that an attacker trying to identify members of the study using an arbitrary dataset might improve their ability to guess study participants at most by a factor of 2.71 times using DP-protected data from the study.

Once again, this is an upper bound. Some attackers might in fact be able to improve their ability to guess study

Figure 15 Differential privacy doesn't prevent data-based deductions that might cause you harm, it only lessens the possibility that those deductions will result from the use of your individual data in forming those deductions.

participants by a factor of 2.71, but other attackers might not experience any improvement.

I took these studies quite personally when they were published, as I'm an Ashkenazi Jew living in the United States. To the best of my knowledge, I wasn't involved in the studies. Nevertheless, people who know that I am an Ashkenazi Jew—you, for instance—might infer that I have a higher chance of colon cancer, schizophrenia, and bipolar disorder after reading a publication based on the study's findings (such as the paragraph above).

I may feel that these studies revealed something private about me, but they do not, at least not according to differential privacy's formal definition of privacy loss. That's because these associations were not discovered using my private data. These studies didn't reveal information about me in particular; they reveal something about Ashkenazi Jewish people in general. But how close are my genetics to the Ashkenazi that the study evaluated? That's another source of my privacy! These studies don't reveal anything about non-Jewish people adopted into Ashkenazi families, for example. Nor are the statistics accurate for a person who has one non-Jewish parent or a non-Jewish grandparent. That's personal information, and that's the kind of information about specific individuals that's missing when we try to apply DP-protected statistics to individuals who were not part of the study group.

People struggle a great deal with the difference between an inference about a person and an inference about a person that is based in part on that person's non-public data. Differential privacy's math says that the latter inference is privacy violating precisely because it requires that person's data. This distinction is fundamental—and underappreciated. If this distinction is ignored, we end up with a privacy definition that would likely prevent the publication of accurate statistics about population subgroups in general. Such definitions would probably prevent the publishing of any statistics at all in short order.

This leaves us with the differential privacy definition of privacy loss—the breakthrough definition based on the intuition that statistics computed without your data cannot violate your privacy, even if you have a relationship to people in the dataset. For some people this definition may not be psychologically or epistemologically satisfying, but it is a definition that has allowed researchers and practitioners around the world to make a tremendous amount of progress on privacy-protecting technologies.

Group Privacy

The formal definition of differential privacy assumes that each record in the database is independent. That's not always the case. In a transaction database, the purchases from the same person might appear several times. There's also the possibility that records from multiple individuals might be correlated; for instance, you might be computing statistics on households and two members of the same household have filled out the same survey.

Group privacy extends differential privacy to handle these situations. A simple formulation of group privacy is to change the definition of neighboring databases: instead of differing in one row, they differ in *k* rows. If *k* is known in advance, then a straightforward approach is to add more noise to achieve the same level of protection, although more sophisticated mechanisms might allow less noise to be added by making additional assumptions about the nature of the groups.

With some data privacy mechanisms, the presence of unknown groups within a set of data being protected can cause the privacy mechanism to fail suddenly. Consider a de-identified dataset of children that includes demographic information about each child's parents and their residences. The de-identification mechanism might work reasonably well for families with one, two, or three children, but if the dataset contains a family with four children, a clever data hacker might be able to perform a database reconstruction and reidentify those four children in short order.

One of the advantages of differential privacy is that the privacy protection degrades gracefully in the presence of groups. Protecting that same dataset with DP, the family with four children might experience more privacy loss—because the sensitivity of some counting queries for those families is four instead of three—but the privacy protection mechanism wouldn't suddenly fail and reveal the unprotected data.

Better Privacy Accounting Can Lower the Privacy Loss!

Surprisingly, it's possible to lower the PLB (ε) of a deployed differential privacy system without making any changes to the system itself, but rather by performing a more detailed analysis of the maximum possible impact that data from a single person can have on the result—that is, placing tighter bounds on a query's sensitivity. That's because the value of epsilon is a function of both the amount of noise that's added and the sensitivity of the calculation. Thus if an analyst can prove that the sensitivity is less than previously thought, then the true value of ε must be lower for a given amount of noise.

This happened in 2016, when researchers at Google reevaluated the company's stochastic gradient descent algorithm.[22] By improving the privacy accounting, the researchers were able to decrease ε for an algorithm that ran in 10,000 steps from $\varepsilon \approx 9.34$ to $\varepsilon \approx 1.26$—all without making any changes to the running code! (In this example, δ was constant at 10^{-5}. The δ tells you that this work was done using approximate differential privacy.)

Developing new techniques to improve and simplify privacy accounting remains an active area of research.[23]

Many people who haven't worked with DP are surprised by these results because they don't fit a mental model of starting with a value of ε and then figuring out what statistics can be generated. But the typical way that DP has been deployed to date is by experimenting with different amounts of noise until the study designer decides on a value that produces statistics that are just barely accurate enough for the given application. Today this approach to tuning differential privacy applications is underappreciated.

Synthetic Microdata

The style of data analysis that I've presented in this book so far is different from the way that data analysis is typically taught in schools.

There once was a time when researchers and even journalists looking for a statistic would seek out published reports that might contain the desired details. Did you want to know the average household income for single-parent households with two children and a grandparent by race living in East Providence, Rhode Island? Your best bet was to search for a federal or state agency that might have published such a report.

That was before the age of big data, fast laptops, and data science programs. Today, people are being trained to answer these questions by downloading a relevant dataset containing de-identified *microdata* and running their own analyses.

The American Community Survey (ACS) is one such dataset, based on questionnaires sent to between 1 and 3 percent of the US population every year. The results are published in two forms: tables containing analyses that the Census Bureau's experts deem likely to be useful for many data users, and microdata that anyone can download and use to perform their own analyses. The tables have more precise geographic resolution, typically at the zip code tabulation area, census tract, and even block group level, but are limited in the details and interactions between variables that they can present. The public use microdata are grouped into Public Use Microdata Areas, which contain at least 100,000 people. That means that East Providence (population 46,929) is grouped with Cranston, Rhode Island, bringing the 2021 population up to 129,429. You might be able to find households that satisfy my criteria above in the public use microdata sample, but you won't be able to tell if those families are in East Providence or Cranston.

Today the US Census Bureau uses a combination of traditional statistical disclosure limitation techniques to protect the ACS, including "swapping, partial synthesis, perturbation, top and bottom coding, [and] coarsening," according to a presentation by Rolando Rodríguez at the US Census Bureau at the ACS Data Users Conference 2023.[24] But the Census Bureau is moving to a new approach that will use differential privacy to create *synthetic microdata*. This approach uses the entire privacy budget once to create a set of statistical models that, in turn, produce a single set of records of synthetic microdata. These new microdata can then be used to create tables or be published in their entirety without creating additional privacy loss. We used a similar approach for the 2020 Census.

Making representative microdata is difficult because they need to capture all the statistical relationships that downstream data users might be interested in analyzing.

In the 2020 Census, there were literally thousands of different queries for each geographic region at every geographic level that were used to create a set of privacy-protected measurements (called the *noisy measurements*). These measurements were then used to create a consistent set of synthetic microdata called the privacy-protected microdata files. Part of this process involved applying invariants, such as a prohibition against allocating people to blocks that contained no housing units. The files were then used to create the tables, but they were also released as their own dataset to the public. This was the first time ever that the Census Bureau had been able to release microdata with such richness and geographic specificity from a decennial census. For sophisticated data users who wanted even more accuracy, the Census Bureau also released the noisy measurements.

When I was hired at the Census Bureau, one of my charges was to make the ACS formally private—that is, make the ACS adopt differential privacy as its protection mechanism. This turned out to be much harder to do for the ACS than for the decennial census for two reasons. First, the ACS asks many more questions than the decennial census. Second, the ACS is a stratified probability sample: the Census Bureau splits the country into many groups (the stratification), randomly picks a sample of households in each group (the probability), and then pursues each household until it responds (that's the mandatory part). This means that some respondent's answers count more than others, which makes computing the sensitivity much harder. In 2020, such computations were beyond DP's state of the art.

Validation and Verification Servers

One of the frustrations—and fears—of using synthetic data is the knowledge that a scientific discovery made with such data might not be real. Instead, the discovery might just be finding an artifact from the process that created the synthetic data. This isn't a risk just for differential privacy: it has the potential of happening any time that data are edited to preserve privacy and then used for scientific research. As researchers Arthur Kennickell and Julia Lane observed in 2006,

> Despite the fact that much empirical economic research is based on public-use data files, the debate on the impact of disclosure protection on data quality has largely been conducted among statisticians and computer scientists. Remarkably, economists have shown very little interest in this subject, which has potentially profound implications for research.[25]

Validation and verification servers are a powerful way to address the problem of false discovery made with data

Figure 16 The block-level data in the 2020 Census data products were generated using differential privacy, with the result that the counts for the number of people on specific blocks may not match other data in the public domain.

that are synthetic or have undergone privacy edits. These servers operate on confidential data. The US Census Bureau operated such a validation server for use with the Survey of Income and Program Participation Synthetic Beta program from 2015 through 2022.

To use either kind of server, an outside researcher would first download synthetic microdata that have the exact same format as the data inside the enclave and then use those data to develop their statistical program. Once their program is ready to go, they submit the program to the statistical agency. The program runs on the server. What happens next depends on the kind of server being used.

Validation servers run the researcher's program and produce a result. This result then undergoes review to ensure that the result does not violate privacy policy, after which the result is released to the scientist. Although it is possible to make results formally private, in practice this would be hard to do because of the difficulty of determining the sensitivity of the overall calculation.

Verification servers compare the result of the researcher's program run on the synthetic data with that of the program run on the confidential data and report if the statistically significant conclusions are still valid. Verification servers provide less opportunity for confidential data to be inadvertently released, but they still result in a small amount of privacy loss that should be accounted for by the organization.

Local Differential Privacy and Randomized Response

Differential privacy can be applied to datasets with data from thousands or even millions of individuals, but it can also be applied to thousands or millions of datasets with data from a single individual. This approach is called *local differential privacy* because the noise can be applied by each individual before they send their data to the data curator. That is, they can apply the noise *locally*—for example, using an app running on each person's smartphone.

The advantage of local differential privacy is that the curator no longer needs to be trusted since the curator is receiving data that have already been privacy protected. This can lower the risk and negative publicity that comes with receiving sensitive user data. The disadvantage is that this local differential privacy adds a tremendous amount of noise to get that privacy protection—so much that it is difficult to perform detailed data analysis.

Local differential privacy is similar to a technique invented in 1965 by Stanley L. Warner called *randomized response*.[26] Warner wanted to ask people sensitive questions like, "Have you had an abortion?" or "Have you tried illegal drugs?" The idea was to let people answer truthfully (which he assumed they wanted to do) while still giving each person the ability to deny their response later. As its name implies, randomized response does this by randomly assigning respondents to answer either the sensitive question or an innocuous one that has a known answer distribution. After the survey is completed, statisticians can compute the distribution of the sensitive questions by taking into account the known distribution of the other answers. The statistics work out for the overall average while preventing an observer from knowing who answered the sensitive questions and who answered the innocuous ones.

The randomized mechanism can be perfectly described using the math of differential privacy, which means that randomized response is differentially private! Indeed, randomized response is a simple form of local differential privacy.

In 2014, Google integrated randomized response into the Chrome web browser to collect sensitive information from users, including the home page that each user had set for their browser and list of applications running on each users' computer. Google planned to collect this information for computer security purposes; the idea was to find which installed programs were correlated with unwanted changes to browser settings. In the interest of transparency, Google published a paper explaining what it was about to do and why the project was ethically permissible before it deployed the technology.[27] The system was called Randomized Aggregatable Privacy-Preserving Ordinal Response (RAPPOR).

To understand why local differential privacy and randomized response needs to add substantially more noise for the same level of privacy protection compared with the trusted curator model, think back to the Laplace mechanism. Recall that the amount of noise that's added for each query is proportional to the influence that one person's data can have on the result. This amount of noise is constant whether the query operates on data from one person or a thousand. This is why DP produces more accurate answers as the number of records involved in the query increases.

Local differential privacy is a series of differential privacy queries where each query is applied to data from a single person. All of these individual noisy query results are then combined into a single answer. The noise protects each user's data from the data curator—for example, Google—as well as the final query results from any downstream data users. But it achieves this protection at a significant cost to accuracy. Local differential privacy is useful for getting answers to one or two sensitive questions, but it's not good for finding interactions between multiple variables.

Differential Privacy for Machine Learning and Federated Learning

Artificial intelligence and machine learning is another area where differential privacy is being applied. Today's systems are trained on large amounts of data to perform tasks—and often the data are highly personal. A major risk in machine learning is that an attacker with access to the model can sometimes learn confidential information about the individuals whose data was used to train the model. Differential privacy can reduce this risk.

For example, consider a system that uses machine learning for face recognition. By analyzing the faces of millions of people, with several samples of each person, the system learns the kinds of differences between faces that signify different people—like eyes that are round versus oval—and the kinds of differences that are inconsequential, such as the direction that a person's eyes might be pointing. It's then possible to use that classifier to determine if two photographs came from the same person or different people. This is the core idea behind nearly every modern system that identifies people based on their faces.

The privacy risk comes from the part of the system that is trained on millions of faces. This part, called a *classifier*, might have millions of different parameters. While some of these parameters encode high-level representations of features like eyes, eyebrows, and noses, others literally memorize specific individual faces. It is sometimes possible for an adversary to mathematically reconstruct one or more of the faces that were used to train the system.[28] This kind of attack is called *training data extraction* or *model inversion*. Metaphorically, I like to say that the attacker turns the model upside down and shakes it until the training data fall out.

When differential privacy is applied to machine learning, statistically controlled noise is injected into the data pipeline as part of the training process. Such approaches, many of them based on Google's "stochastic gradient descent with differentially private updates," can limit the probability of extracting training data from a classifier.[29] Recently, DP has been applied to large language models, where it can similarly decrease the possibility of extracting sensitive training data.[30]

Protecting machine learning models from attack is especially important when these models are deployed onto consumer smartphones or video cameras. Since these devices are physically under the control of end users, it can be hard, if not impossible, to prevent capable malicious users from getting direct access to models and attempting to reverse engineer them. Without the protection of differential privacy, many applications may not be realized because of the potential privacy risk.

Major Figures and Key Thinkers

Differential privacy was invented in 2006, but it built on decades of work by others in the field of statistical disclosure control. Here I present some of the critical people

Figure 17 Training data extraction is like turning a model upside down and shaking it until the training data fall out.

who helped develop prepare the world for DP, as well as those who have contributed to DP's theory and practice.

Before Differential Privacy

The groundwork for today's data-rich world was laid at statistical agencies, where fundamental techniques for record linkage, imputation, and disclosure limitation were first developed. As its name implies, record linkage finds connections between records in different datasets. Imputation fills in holes left by missing data. Disclosure limitation makes it possible to use and publish statistics while limiting the impact on privacy.

- *Tore Dalenius* (1917–2002) was an employee of Statistics Sweden. Much of his early work was published in the Swedish language journal *Statistik Tidskrift*, the predecessor to today's *Journal of Official Statistics*. He proposed a formal methodology for statistical disclosure control in 1977.[31]

- *Ivan Fellegi* escaped Hungary after the failed Hungarian Revolution of 1956, earned an MSc and PhD from Carleton University, joined Statistics Canada (then the Dominion Bureau of Statistics), and rose through the ranks to become Canada's chief statistician from 1985 to 2008. Fellegi's personal experience with governmental repression drove much of his interest in statistical confidentiality. "The concern is real and the danger is also real," he wrote in 1972.[32]

- *Latanya Sweeney* gained fame when she showed in 1997 that Governor Weld's medical records could be found in a public dataset—one of the earliest public demonstrations that the de-identification approaches developed between the 1960s and 1990s were not sufficient for protecting privacy in a world of widely available public data. Now a professor at Harvard University, Sweeney and her students have successfully reidentified individuals in many de-identified data releases.

- *Irit Dinur* coauthored the 2003 paper with Kobbi Nissim that resulted in what is now called the fundamental law of information recovery and led to the creation of differential privacy.[33] Both Dinur and Nissim were living in Princeton, New Jersey, at the time, although they were affiliated with different institutions. While Nissim continued working in privacy, for Dinur, privacy was just a diversion; her primary research is in the fields of coding theory and probabilistic proof checking, for which she was awarded the Erdős Prize in 2012 and the Gödel Prize in 2019.

- *Stephen Fienberg* (1942–2016) was a preeminent statistician and professor of statistics and social science at Carnegie Mellon University. He was the founding editor and editor in chief of the *Journal of Privacy and Confidentiality*, which he cofounded with Cynthia Dwork.

  In summer 2005, Fienberg and Dwork brought cryptographers and statisticians together in the hillside town of Bertinoro, Italy, at a workshop exploring approaches for formalizing statistical disclosure control.[34] Later, Dwork described the *Fienberg problem*, which is the privacy leakage resulting when a well-intentioned statistician looks at confidential data, and based on the unprotected data, decides which statistics to release. It turns out that this decision process itself leaks confidential information that is unaccounted for in traditional disclosure avoidance approaches and can be utilized by an adversary to learn details of confidential data that should be protected.[35]

- *Helen Nissenbaum* is a philosopher and professor of information science at Cornell Tech. She is best known for her privacy theory of *contextual integrity*, which is a qualitative approach for evaluating the collection and use of confidential information. These approaches do not directly map to differential privacy, but they can be useful when deciding how to use differential privacy and setting PLBs.

DP's Inventors

DP was invented by Dwork, McSherry, Nissim, and Smith. By convention, their names always appear in alphabetical order, matching the order of their 2006 paper, "Calibrating Noise to Sensitivity in Private Data Analysis." For this work, the authors were awarded the Theory of Cryptography Conference Test of Time Award in 2016, Gödel Prize in 2017, and Association for Computing Machinery's Paris Kanellakis Award in 2021.

The 2006 paper is the correct reference for differential privacy, and not Dwork's single-author paper from that same year, which is also called "Differential Privacy" and further develops the ideas in the joint paper.

- *Cynthia Dwork* was a research scientist at MSR from 2001 until 2023. In 2017, she joined Harvard University's faculty as a Gordon McKay Professor of Computer Science at the Harvard Paulson School of Engineering and Applied Sciences.

- *Frank McSherry* was also a researcher at MSR when he coinvented DP. After the paper was published, McSherry developed Privacy Integrated Queries, the first interactive differentially private data analysis system.[36] McSherry took an early lead in directly arguing with differential privacy critics.[37] McSherry and Kunal Talwar developed the DP *exponential mechanism*, for which they were awarded the 2009 PET Award for Outstanding Research in Privacy Enhancing Technologies.[38] Today McSherry is the chief scientist at Materialize, a database software company.

- *Kobbi Nissim* coinvented DP when he was a postdoc at MSR; today he is a professor of computer science at Georgetown University. In 2003, he coauthored "Revealing Information while Preserving Privacy" with

Irit Dinur; this joint work was awarded the 2013 ACM PODS Alberto O. Mendelzon Test of Time Award.

• *Adam D. Smith* was employed at MSR in summer 2005, but had started a fellowship at the Weizmann Institute of Science in Israel when he coauthored the 2006 paper. He then was a visiting scientist at both the Massachusetts Institute of Technology and the University of California at Los Angeles before joining the faculty of Pennsylvania State University. Smith moved to Boston University in 2017. In addition to his awards on differential privacy, he shared the 2019 Eurocrypt Test of Time Award with Yevgeniy Dodis and Leonid Reyzin for their work on extracting strong keys from biometrics and other noisy data.[39]

DP's Scientists

One of the marks of DP's acceptance and generality is that many research scientists have been able to advance the theory in many different directions. This did not happen with any previous mathematical approach for statistical disclosure control.

• *Daniel Kifer* was the early scientific lead on the development of the TopDownAlgorithm used in the 2020 Census, along with Robert Ashmead, Ryan Cumings-Menon, Phil Leclerc, and Brett Moran. Pavel Zhuravlev later joined the core team (as did I). Kifer is a professor of computer science at Penn State University.

- *Ilya Mironov* discovered in 2012 that all implementations of differential privacy up to that point leaked private information because the DP theory assumed that numbers were continuous, but the actual implementations used binary-encoded floating point numbers. "Unlike its mathematical abstraction, the textbook sampling procedure results in a porous distribution over double-precision numbers that allows one to breach differential privacy with just a few queries into the mechanism," he wrote.[40] This problem is now mostly addressed.

    Mironov also invented Rényi differential privacy and worked with others at Google Research to develop the "amplification by shuffling" mechanism, which improves the privacy/utility trade-off of local differential privacy by combining it with a mixing network to provide an additional layer of uncertainty.[41] This makes it possible to decrease the amount of noise used, although it does so by introducing a weaker threat model and corresponding weaker privacy guarantee called *shuffle privacy*.

- *Aleksandar Nikolov* is a professor at the University of Toronto, and holds the Canada Research Chair in Algorithms and Private Data Analysis. He has contributed to developing algorithms that allow for private analysis of histograms and other kinds of statistical queries.

• *Kunal Talwar* coinvented the exponential mechanism with McSherry and participated in the development of Google's amplification by shuffling mechanism.[42] Since 2019, Talwar has worked as a research scientist at Apple.

• *Aaron Roth*, a professor at the University of Pennsylvania, is known for his extensive work in algorithmic game theory and coauthoring (with Dwork) *The Algorithmic Foundations of Differential Privacy*, the first differential privacy textbook.[43] Like the book you are reading, Roth and Dwork's book can be freely downloaded over the internet thanks to its open-access license.

• *Vitaly Shmatikov* is a professor of computer science at Cornell Tech. Shmatikov and his students have published many breakthrough papers that have significantly contributed to our understanding of privacy and the impact of differential privacy, mostly by showing that approaches for preserving privacy rarely work as well as their creators think. With Arvind Narayanan (now a professor at Princeton), Shmatikov demonstrated that many data records in a large "anonymous" dataset released by Netflix could be reidentified.[44] With Reza Shokri (now at the National University of Singapore), Marco Stronati (now a research scientist at Matter Labs), and Congzheng Song (now at Apple), he quantified the privacy risk of membership inference attacks, showing them to be a better measure of risk when using private data to train a machine learning system than database reconstruction or reidentification.[45] With Eugene Bagdasaryan (now a professor at the University of Massachusetts at Amherst) and Omid Poursaeed (now a research scientist at Meta AI), he found that differential privacy applied to machine learning tends to drop the accuracy for underrepresented classes and subgroups far faster than for those in the majority.[46] This has important implications for fairness and bias, although another way of thinking about these results is that DP is doing what it is supposed to do, which is protecting both members of the dominant classes and outliers. With his students, Shmatikov won the Caspar Bowden PET Award for Outstanding Research in Privacy Enhancing Technologies in 2008, 2014, and 2018, and has received numerous other accolades.

DP Deployments

Here I briefly recount the major deployments to date of differential privacy. You will find an up-to-date list in the Wikipedia article "Implementations of Differentially Private Analyses," which I and others periodically update.

Figure 18 Differential privacy can help level the field between data elites and the rest of us.

OnTheMap (US Census Bureau). The world's first production deployment of differential privacy was the US Census Bureau's OnTheMap interactive map.[47] Accessible online at https://onthemap.ces.census.gov, the map invites users to select or draw a geographic boundary of any size. The system can then provide information about the people who live in the area, such as their jobs and salaries. These data can be sliced many ways—for example, by worker race and ethnicity, education, and sex. The system can also reveal commuting patterns. Using OnTheMap, a city can determine how many of its residents work in the suburbs—or an adjoining state.

OnTheMap's data come from many sources, and there are so many ways to slice and display them that the people who created the map in the early 2000s realized that there would be a significant disclosure risk if they did not address privacy issues from the start. The team decided to create synthetic data using noise infusion and then use those synthetic data to power the map visualization. Differential privacy was invented while the team was working on the project, so the DP noise approaches were slotted in.

OnTheMap was created by Cornell University professors John Abowd and Johannes Gehrke, Gehrke's students Ashwin Machanavajjhala and Daniel Kifer, and Abowd's longtime collaborator Lars Vilhuber. The experience proved to be formative, with Abowd returning to the Census Bureau to become its chief scientist in 2016 where he oversaw the adoption of differential privacy, Kifer returning on sabbatical to design the TopDown Algorithm that would power the privacy mechanism for the 2020 Census redistricting products, and Machanavajjhala becoming a cofounder of Tumult Labs, which would create a commercial privacy-protecting database that created detailed demographic and housing statistics for the 2020 Census. In 2024, Abowd received the Association for Computing Machinery's Policy Award "for transformative work in modernizing the US Census Bureau's processing and dissemination of census, survey and administrative data, exemplifying privacy-aware management of government-collected data."

Google Chrome. Google introduced differential privacy into Chrome, the world's most widely used web browser, to collect information from users including the home page that each user had set for their web browser and a list of all processes running on the computer that was also running Chrome. One application for these data is to determine which computers are infected with malware and correlate this information with the websites that infected users have visited. Chrome always had access to such information, of course, but the information was sensitive, and for that reason, Google had not used Chrome to collect the data from each person's computer and send it to Google's servers.

Local differential privacy gave Google a way to collect this information and apply privacy protection directly on each end user's computer, so that Google would never get the raw sensitive information. Google developed the technology and presented it at one of the world's top academic computer security conferences.[48] The researchers were Úlfar Erlingsson and Vasyl Pihur at Google, joined by Aleksandra Korolova at the University of Southern California. The system, as mentioned earlier, was called RAPPOR.

Google's 2014 RAPPOR paper had a significant and continuing impact on differential privacy. The fact that Google had deployed differential privacy in Chrome brought instant credibility to DP, which was then only eight years old. But it also caused many people to conflate DP and randomized response.

Google ultimately grew so dissatisfied with RAPPOR that it developed a secondary privacy-protecting mechanism called *amplification by shuffling*, which allowed Google to significantly decrease the amount of noise added to each user's data.[49] Contributors to that project included Erlingsson, Vitaly Feldman, Mironov, Ananth Raghunathan, Talwar, and Abhradeep Thakurta.

Private federated learning, Google's second major contribution to the field of DP, lets Google train complex machine learning models using private data from Google users, without ever having confidential data leave the end point where the data are stored. Instead, the model is split up into tiny pieces, with each piece being sent to a different end point. Traditional federated learning can be used to train machine learning systems from data on individual user cell phones or medical records scattered across many facilities, without the need to bring all of the data together to one place. With private federated learning, noise is systematically added as part of the federated learning process. Google's TensorFlow Privacy and its FL-DP, DP-SGD, and DP-FedAvg algorithms all provide versions of federated learning that use differential privacy.

Consumer operating systems from Apple and Microsoft. Both Apple and Microsoft use collect differentially private statistics from end users with their consumer operating systems. Apple uses differential privacy to discover new words for autocorrect, improve emoji prediction, and find websites that use excessive CPU or memory within the Safari web browser.[50] Microsoft uses differential privacy to collect telemetry data such as daily app usage statistics from Windows installations.[51]

Uber. In 2014, *Forbes* published an article claiming that the ride-sharing company Uber had implemented a system called "God View" to let Uber's senior management monitor users by name in real time.[52] The God View allegations resulted in a 2016 settlement between Uber and the New York attorney general, and was reportedly a key issue in the company's 2017 settlement with the US Federal Trade Commission.[53] Also in 2017, Uber announced that it had developed a system that used differential privacy to allow analysts within the company to search databases in a privacy-preserving way.[54] The system lets analysts conduct queries on real customer data while reducing the risk that those same analysts might be using Uber's datasets to monitor the movements of specific people. Uber released the source code for its system in 2017. Similar technology is now being commercialized under the trademark PrivateSQL by Oasis Labs.

The 2020 US Census of Population and Housing. The most complex deployment of DP to date was the use of DP for the twenty-fourth US Census, the 2020 Census of Population and Housing. This effort involved literally hundreds of people, including dozens of statisticians, scientists, engineers, attorneys, and community engagement professionals at the US Census Bureau, along with dozens of academics who reviewed and commented on the preliminary statistical products produced by the bureau as well as countless journalists, community organizers, and academics who participated in the public discussion about the decision to move from traditional disclosure limitation approaches to differential privacy.

DP's Start-ups and Investors

Most of DP's first ten years were spent in the research hallways of major corporations, large government agencies, and the world's top universities. This reflects the fact that it is only these kinds of organizations that can afford to spend the large sums of money required to invent and nurture this kind of specialized scientific product.

For example, OpenDP is a project hosted at Harvard University that is building open-source tools for differential privacy. The project's eponymous tool kit is rich library of differential privacy mechanisms and algorithms written in the Rust programming language, and based on the SmartNoise library, which was created in a collaboration between Harvard and Microsoft. OpenDP is headed by Harvard professors Gary King and Salil Vadhan. It has received substantial funding from the Sloan Foundation, which also provided funding for this book.

But there's also money to be made with DP: many organizations have large troves of personal data that they would like to use for research or commercial purposes. DP's ability to produce a worst-case mathematical bound on privacy loss gives these organizations a way to achieve their goals while explicitly limiting the risk that can result from the use of private information.

Companies pursuing differential privacy fall into two categories:

- Companies including LeapYear, Oasis Labs, Privitar, and Tumult have developed databases that hold confidential data, accept queries, and apply differential privacy to the results. These databases can be used interactively, limiting the potential for data snooping on the part of analysts; they can also be used to produce protected tables or even synthetic microdata.

• Companies including Duality Technologies and Galois have combined differential privacy with other privacy-enhancing technologies such as secure multiparty computation (MPC) and HE so that the outputs of these techniques are themselves privacy protected.

DP deployments will likely require specialty software and trained consultants for another five or ten years, replicating the experience and timeline of public key cryptography, another math-based approach for protecting privacy. By the mid-2030s, however, DP should be widely understood, taught in high schools, and easy to deploy.

2

DIFFERENTIAL PRIVACY ISSUES

Today DP is seen by many as the gold standard in privacy protection, but it's also seen as a technology that is difficult to use and imposes restrictions that might not be necessary for real-world applications. I believe that this is a reflection not of the underlying math but instead the fact that DP is a relative newcomer; in 2026, DP will be only twenty years old! DP will probably be less controversial when it is used in the 2030 Census, and by 2036—three decades from its invention—it should be widely accepted.

Thirty years from scientific invention to widespread acceptance might seem like a long time for a technology to get deployed, but it's similar to the adoption curve for public key cryptography, another math-based privacy technology that migrated from the world of theoretical computer science to practical day-to-day applications. Although public key cryptography's three critical inventions were made in the 1970s, it wasn't until 2014—thirty-seven years after public key infrastructure's invention—that there was widespread agreement that *everything* sent over the internet should be encrypted, rather than simply passwords and credit cards.[1]

Questions and Conflicts

DP has significant advantages over legacy approaches for statistical disclosure limitation:

- DP provides formal, mathematical guarantees that the privacy impact of any given data release will be limited to a knowable PLB.
- This PLB can be adjusted as necessary to balance requirements of accuracy and privacy protection.
- DP composes, making it possible to determine the cumulative privacy loss caused by many uses of the same confidential data.
- DP supplies a worst-case analysis that does not depend on the attacker's capabilities or auxiliary knowledge.

- The DP guarantee for a specific data release does not degrade over time as computers get faster and more auxiliary data becomes available.
- DP's guarantee gracefully degrades if the underlying assumptions of a particular use aren't rigorously followed—for example, if a person's private information appears twice in a database instead of once or if two records are correlated.
- DP has no secret parameters. The DP pipeline can be described, and all the privacy-related source code can be published. In this manner, DP allows for transparency in making statistics.

At the same time, DP has many limitations. Let's explore them in turn.

The Inherent Conflict between Privacy and Accuracy

Differential privacy forces us to confront the reality that every use of confidential data inevitably results in some privacy leakage. There is no free lunch in data privacy! This is a frustrating situation, but it reflects the true nature of the world. It is the statistical analog of physicist Werner Karl Heisenberg's uncertainty principle, which holds that the more accurately we know the position of a particle, the less accurately we know the combination of its mass, speed, and direction of travel.

Differential privacy forces us to confront the reality that every use of confidential data inevitably results in some privacy leakage. There is no free lunch in data privacy!

The Question of "Worst-Case" versus "Average-Case" Analysis

Differential privacy gives a worst-case measure of privacy loss: follow the rules and crunch the math, and you get a guaranteed maximum amount of privacy loss that your data subjects will suffer when your statistics are pitted against a data hacker with access to an unlimited amount of computer power and auxiliary data.

This may be a comforting degree of privacy protection, but the idea of a data hacker with all the world's data (except the data you are protecting) and an unlimited amount of computer power is unrealistic. Sadly, pure differential privacy provides no guidance as to what kinds of privacy protections are needed against more realistic data hackers.

Many DP critics have the intuition that less noise might be required to defend against more realistic opponents. They ask, *If we are only trying to defend against an adversary with a few dozen workstations and a year's worth of data from the* Financial Times *and US Census Bureau, can we get by with less noise? Please?*

Frustratingly, the answer to this question is, *We don't know!* It's mathematically easy to model a computationally unbounded adversary that has unlimited access to data, but it's difficult to model an adversary with a realistic amount of computer power and access to specific datasets. And what happens if that assumption turns out to be wrong?

Data Exploration

The simple DP model provides for processing a single confidential dataset to make a static set of statistics that will be published and widely distributed. This model is less successful when trying to protect confidential data from a data analyst making multiple interactive queries. Should the data analyst have a single privacy budget, or should there be a new privacy budget every day? Should the privacy budget only be expended when the analyst publishes a table and not when they are simply doing exploratory work? If there are five data analysts, should they each have their own budget? What happens when new data arrive next month?

Different organizations have answered these questions in different ways, and while there is a mathematically correct answer, it is so draconian that few organizations want to work under the true DP rules.

A true, faithful respect of the DP rules holds that *every* use of confidential data must be accounted for in any precise tally. This means that if five analysts make 100 queries a day on the confidential data for a week, and each query has an $\varepsilon = 1$, then the epsilons just add up. This means that the total privacy loss is $\varepsilon = 5 \times 100 \times 5 = 2{,}500$, which is a laughably high value of ε (although we could get a much lower value of ε by using zCDP and an advanced composition method).

But remember, this computation is for the worst-case analysis, and here the worst case would be that all the analysts collude to craft a set of queries with the highest probability of revealing the confidential data. Another possibility is that each of the analysts is working on their own project, and at the end of the week only one or two results actually get published, with the rest being thrown away and the analysts forgetting everything else. Thus the total effective privacy loss might just be a value of 1 or 2.

There is a middle ground, but it's this middle ground that makes it harder to know what is really going on in this example. It's likely that some of the analysts use their queries at the beginning of the week to get a better understanding of the data so that they can create more specific queries that do a better job of coming up with useful results later in the week. This kind of leakage is similar to the problem of overfitting in machine learning, except that instead of overfitting a classifier to the data, the repeated analysis overfits privacy safeguard. This risk is especially likely when developing DP mechanisms tuned to a specific dataset.

We addressed this issue in the decennial census by developing our software using data from the 2010 Census. The thousands of runs we did with the 2010 data did not count against the PLB for 2020.

Another approach for performing adaptive data analysis was described by Dwork and her coauthors in a 2015 article that appeared in *Science*, one of the world's most prestigious scientific journals.[2]

Semantic Interpretations

Perhaps the deepest conflict within the world of differential privacy researchers and practitioners is the most fundamental: What does differential privacy's measure of privacy loss mean? What is the true meaning of ε? What is the meaning of δ? What is the true impact of doubling the PLB? What is the bad thing that happens if epsilon is too big? This is the differential privacy conflict over *semantics*.

Just like the math of quantum physics, there are many ways to interpret the math of differential privacy. But whereas quantum physics is ultimately grounded in observations of the world around us, differential privacy is based on a philosophical examination of analyzing confidential data and the impact of auxiliary information. Policymakers might prefer ε and δ to be an absolute measures of privacy protection, allowing them to state unequivocally that a data release is privacy neutral or will have only minimal impact on the privacy of the people on whose data the release is based. Auxiliary information foils this desire.[3]

Many differential privacy researchers have adopted the Bayesian interpretation of differential privacy, meaning that the privacy impact of a data release depends entirely on an adversary's prior knowledge.[4] This is not satisfying from a public policy point of view because people implementing policy need to supply concrete assurances that their actions are within the legal requirements.

Thus a significant effort that was part of the 2020 Census DP was to develop a scientific paper that presented something closer to an absolute guarantee of privacy using the DP mathematics.[5]

DP Critics

Differential privacy has been surprisingly controversial. Most of the attacks have been aimed against the use of differential privacy in the US 2020 Census, although some critics have contested the very idea that differential privacy is an appropriate statistical protection for confidential data. The criticisms have come from many directions. The remainder of this section discusses each of these arguments in turn. Because attacks over the use of DP have occasionally devolved into each side questioning the other's motivations, I will refrain from mentioning specific DP critics by name, and do not include any references. Instead, I'll simply summarize the critiques and provide a response.

> *It's a nice theory, but it doesn't work in practice.*
>
> *Traditional techniques do an adequate job protecting privacy, and without significant impact on data accuracy.*

Figure 19 Arguments against differential privacy.

Figure 20 More arguments against differential privacy.

Differential privacy starts with a mathematical definition of privacy loss that is based on an intuitive understanding of privacy: your privacy isn't violated if your confidential data are not used to produce a statistic.

These assertions typically come from statisticians and social scientists with significant experience in traditional disclosure avoidance techniques. These professionals argue that legacy techniques are sufficient (if not better) than DP for protecting privacy.

The most obvious response is what Nissim said to me back in 2016: "What, you don't believe the math?"

The fundamental difference between differential privacy and legacy techniques is that differential privacy starts with a mathematical definition of privacy loss that is based on an intuitive understanding of privacy: your privacy isn't violated if your confidential data are not used to produce a statistic. That intuition produces the DP definition, which is what DP mechanisms implement.

Critics can (and do) contend that DP's definition doesn't make sense, but none have proposed an alternative one that is mathematically rigorous. Indeed, the history of statistical disclosure limitation before DP was that statisticians started with a mechanism, such as swapping or suppression, and then argued that there was no practical way for a data hacker to reverse the mechanism and recover the confidential data. Such reasoning inevitably failed to take into account auxiliary data that might be available to the data hacker now or in the future. The field of cryptography went through a similar evolution from ad hoc to formal methods at the end of the twentieth century.

The math is too complicated to understand! True transparency requires something that can be explained to someone with a high school education.

While proponents of DP argue that their approach provides for transparency, critics sometimes maintain that because the math is so complicated, the transparency is illusionary. This contention mischaracterizes what is meant by transparency.

In scientific discourse, the word *transparency* doesn't mean that something can be understood by anyone with a sixth-grade education in math. Rather, it means that every aspect of how something is designed, implemented, or used can be learned in principle by someone who is willing to spend the time to master the intellectual complexity of the subject at hand. Hence the design and certification of skyscrapers is transparent because building codes are based on principles of physics and engineering that are readily available. On the other hand, the design and certification of nuclear weapons is not transparent because much of that information is classified.

Likewise, the legacy disclosure avoidance methods that the US Census Bureau used for the 2010 Census are also secret: the bureau did not divulge the details or publish its code, and both are shielded from Freedom of Information lawsuits. The same goes for today's American Community Survey, which has not yet transitioned to formal privacy methods. In contrast, the bureau published the software that implemented the disclosure avoidance algorithm of the 2020 Census, and additionally engaged experts and community members regarding the choice of the PLB.

> *Statistical agencies should not be adding intentional errors to published statistics that will be used by researchers and policymakers. Science is more important than privacy! Differential privacy will kill babies!*

Many economists and social scientists were dismayed to learn that the Census Bureau planned to use differential privacy to add intentional errors to published data to satisfy the bureau's legal obligations under Title 13 of the US Code. Surprisingly, we discovered that some of the people who complained didn't know that disclosure avoidance techniques had been applied to the decennial census since 1960.

Some medical researchers have voiced concern that the noise added by differential privacy might result in medical models containing errors as well. A 2014 study published at the prestigious USENIX Security Symposium found that at values of epsilon "effective at *preventing attacks, patients would be exposed to increased risk of stroke, bleeding events, and mortality*" using personalized warfarin-dosing models created with differential privacy. "We conclude that *current* DP mechanisms do not simultaneously improve genomic privacy while retaining desirable clinical efficacy, highlighting the need for new mechanisms," the study noted.[6] Fortunately, improved mechanisms have been developed in the intervening years.

The fear that intentional errors might result in incorrect scientific and policy decisions is a real, but the chance of this is less using differential privacy than with legacy disclosure avoidance techniques. That's because the impact of differential privacy can be computed on statistical outputs—something that is only possible because of DP's transparency.

Anyway, all of these data are public already, so why bother protecting them?

Some critics argued that there was no need to apply privacy protections to the 2020 Census because the data were already available from public sources. This assertion is logically inconsistent because if the data were in fact publicly available, then researchers could use those public sources rather than the data published by the Census Bureau!

In truth, the data published by the Census Bureau are not available from any other source, nor are there high-fidelity surrogates. Especially important is the coverage of US rural areas, people under eighteen, the self-reported race data, and the occupancy of group quarters. Many of the people who claimed that the data were already public and releasing them wouldn't cause harm were those in positions of power, generalizing from their own experiences. Revealing the number of people living in a suburban upper-middle-class house might not be a big deal to those residents, but revealing that there are six people living in a New York City subsidized apartment that is only approved for four people could result in their eviction.

Differential privacy erases the contribution of small minority communities in public statistics.

Differential privacy adds noise to statistics, which means that the noise required for effectively protecting members of the majority can dominate or even erase the contributions of small, minority communities. Here the word *minority* does not mean a racial minority but instead any small group that is present in a larger population. DP will hide the existence of five White people in a community of five thousand nonwhites just as effectively as five Indigenous people in a community that is predominantly White and Black or African American. Because DP's noise is additive, it has significantly more impact on a subgroup of five people than a subgroup of five thousand in a population of ten thousand.

DP doesn't need to have this impact; mechanisms could be developed that provide different PLBs to different

Figure 21 When people generalize about data from their own experiences, they're usually wrong.

Table 3 Per Attribute Epsilons for the 2020 Census Redistricting Files

Attribute	Epsilon allocation for all such questions using sequential composition
Housing type (HHGQ)	7.24
Voting age (VOTINGAGE)	7.57
Hispanic? (HISPANIC)	10.04
Census race (CENRACE)	10.08
Occupancy status (H1)	2.47
Total epsilon	37.40

Source: US Census Bureau, *Disclosure Avoidance for the 2020 Census: An Introduction* (Washington, DC: US Government Printing Office, November 2021), https://www2.census.gov/library/publications/decennial/2020/2020-census-disclosure-avoidance-handbook.pdf.

communities. For example, in the 2020 Census, a total PLB of 10.08 was allocated to all questions involving race, but only a budget of 2.47 was allocated to questions involving occupancy status (see table 3).[7] Just as differential privacy allowed the Census Bureau to release data that had more accuracy devoted to each person's race than to their voting age or the precise kind of housing unit in which they were living, so too could some future statistical agency devote more PLB to queries that counted the number of people of a specific race or subpopulation. The difficult part of such a project would not be implementing the mechanism; it would be deciding which groups would get more privacy and which groups would get less. That's a process that would likely be politically challenging, but not technically difficult.

Differential Privacy Technical Limitations

Differential privacy is less than two decades old, so it's not surprising that it is still quite limited. In some cases, these limitations reflect the current state of the art, while in others we suspect that these limitations are inherent in the way that technical communities have defined the problems that they want to solve. Here I'll look at a few limitations and provide my opinion as to whether these limitations are inherent.

Time Series

Much of the interesting data about people have an element of time. A *time series* is a special kind of dataset where the data represents measurements on the same entities taken at multiple points in time.

Consider the dataset from the first chapter, in which nine students get 100 percent on an exam and the tenth gets 80 percent. This dataset could become a time series when we receive the grades from the second and third test in the course.

Today differential privacy works well for reporting statistics separately for each test event, but not if we wish to link the results and release them incrementally. If the second and third tests have the same distribution of nine students with 100 percent and one student with an 80 percent grade, there is a big difference if it is always the same student getting the 80 percent or a different student each time.

Time series are problematic for DP because each subsequent analysis of the confidential data impacts the privacy loss, and linking the data from the second and third tests with the first is an additional use of the first dataset. This is another area of active research.[8]

Graph Data and Linked Data

In data science, the word *graph* means a collection of relationships between points (called nodes or vertices) and the lines that connect them (called edges or links). Graphs can be used to represent highway maps (the cites are the nodes; the highways are the edges) or social networks (each person is a node; friendship is indicated by an edge).

Differential privacy is remarkably bad at protecting graph data because of the outsized impact that a single node can have on the entire social network, depending on where that node exists and to whom it is connected. Mathematically, this is because modifying one node can potentially modify *every other node in the network*. Consider the social network graph for William Shakespeare's tragedy *Romeo and Juliet*: when Romeo dies, the network is partitioned; no such partitioning happens when Tybalt kills Mercutio. These characteristics make graphs both interesting to study and difficult to protect with differential privacy.

Linked data are another special kind of data that differential privacy is not good at protecting, at least not yet. For decades, the decennial census has collected information about both population and housing, and these two datasets are linked together, in that each person is linked to the place where they reside. For the 2020 Census, however, the bureau published distinct protected datasets that are not linked—one dataset for population, and the other for housing. By design, researchers could not link these two protected datasets.

Narrative Text and Other Kinds of Nontabular Data

Differential privacy was designed to protect the kind of tabular data that is part of many statistical datasets. Increasingly, data scientists are working with narrative text and other kinds of nontabular data.

For example, here is an excerpt from a bonding report prepared by the Retail Credit Company and submitted to the House Subcommittee on Invasion of Privacy in the 1960s. We are told that the subject of the report was a retired army lieutenant colonel:

> He was known to be a rather wild tempered, unreasonable and uncouth person who abused his rank and wasn't considered a well adjusted person. He was known to roam the reservation at Ft. Hood and shoot cattle belonging to ranchers who had leased the grazing land from the army.[9]

There aren't many approaches for protecting this kind of information with differential privacy. We can't simply assign each word a number based on its position in the dictionary and then apply Laplace noise based on some value of ε; changing words at random is almost certain to produce nonsense.

A possible approach would be to extract some kind of semantic meaning from the paragraph and create a vector representation of different human characteristics, and then apply DP to that vector. Yet another approach might be to give up on making this paragraph differentially private, and instead focus on protecting statistical metrics based on an analysis of this and other text in the database. Needless to say, protecting text with differential privacy is still very much a research problem and nowhere near ready for deployment.

Photographs

Researchers have also explored how differential privacy might be used to protect photographs of people on streets, such as those that appear on maps in Google Street View.

Google first launched Street View in May 2007, and the application soon came under criticism for showing people's faces and views inside their houses.[10] Google ultimately responded in 2009 by blurring faces.

Face blurring is a de-identification technology that does not have any mathematical basis for its effectiveness. It relies on the assumption that it's hard to identify a person if you can't see their face. Many people, however, can be identified by friends and associates even if their face can't be seen because of the way that the person dresses, the shape of their body, how they stand, a characteristic disability, or where the photograph was taken.

There are other approaches for protecting the identity of individuals in photographs. By using advanced artificial intelligence algorithms as opposed to face blurring, the person's image can be removed from a scene and replaced by another person. Using differential privacy, we could design a system that parameterized thousands of aspects of a person's face and then applied noise, making it possible to trade off accuracy and privacy loss. At low values of privacy loss, there might be a high probability of significant changes to a person's sex, age, race, height, and so on. At high values of privacy loss, these metrics might look closer to the original, but face recognition systems might still report only the possibility of a match.

We can imagine that Google could simply use artificial intelligence to remove all the people entirely from a street scene, but then its Street View would have the eerie feeling of driving through a ghost town. Removing racial characteristics would also remove ethnic authenticity from street scenes. The same goes for age and gender; consider two versions of the same Street View scene, one filled with old men, the second filled with young children of both sexes. There is a compelling reason to maintain some accuracy of a person's demographic characteristics on Google Street View, but how much? These are some of the reasons that de-identification of photographs, movies, audio recordings, and other kinds of multimedia is an active research area.

Medical Records

Medical records are a rich combination of time series information, laboratory test results, textual narratives in the form of physician notes, and many images such as X-rays, CT scans, ultrasounds, and conventional photographs. Medical records are used for both treatment and research, but research using medical records is severely limited by privacy concerns.

Today it is possible to use differential privacy to protect tabular medical records that are used in simple research studies. Looking to the future, we know that many more medical discoveries will be made as we combine information in medical records with other information such as the patient's diet, job, commute, and genome. Differential privacy is not yet sufficiently advanced to protect this kind of multimodal research, but researchers are hard at work developing approaches that might be used in the coming years.

Secure Data Enclaves

Secure data enclaves are places where people can work with confidential data. What makes them "secure" is that you can't get the data out without going through some sort of formal review process that assures the privacy mechanism will be properly implemented for every data release. These enclaves are sometimes called *trusted research environments* or *secure research environments*.

Federal Statistical Research Data Centers form a network of secure data enclaves operated by a partnership between the US government and several research institutions that let vetted researchers work with confidential US government data. These centers have sensitive data from multiple sources, including the US Census Bureau, US Internal Revenue Service, and Department of Health and Human Services. Federal Statistical Research Data Centers are physical rooms where researchers can work (ideally, leaving their cell phones on the outside), but increasingly they are virtual environments that a researcher can access from their own office using special software.

It is also possible to create a data enclave entirely within advanced microprocessors in a special secure processing area called a *trusted execution environment* (TEE). In these cases, the data are made available to researchers in an encrypted form that can only be decrypted inside the microprocessor by a specific program that has been vetted and approved by the data owner. The program typically performs some kind of statistical manipulation, but will not allow the decrypted confidential data to be exported from the data enclave.

Policy Shortcomings

In addition to the technical limitations, there are many policy shortcomings that limit the deployment of differential privacy.

People Implementing Privacy Policies Lack Tools and Experience for Setting the PLBs

One of the core principles of DP is that policy should decide the trade-off between accuracy and privacy loss. But the research community has been slow to create tools that can help people implementing policy to set the multiplicity of parameters and metaphoric tuning dials that DP gives them.

Figure 22 Federal Statistical Research Data Centers rely on up-to-date privacy technology to restrict confidential data to people who are authorized, unlike the data center pictured above.

People Need Help Understanding What "Privacy Loss" Means

Differential privacy provides a quantitative way to measure the maximum possible privacy loss that an individual suffers as a result of any given data release, but it doesn't provide us with tools for understanding what that means or putting the loss into context. Privacy loss isn't like smoking or radiation exposure, where individuals who suffer more than a certain amount are highly likely to experience bad outcomes. But unnecessary privacy loss really is something that's better avoided; it's not the same as people who won't step on cracks between sidewalk tiles to avoid breaking their mother's back.

Differential privacy's concept of privacy loss doesn't take into account the actual risk or harm that any given data release might have. For many people, it's far worse to leak their income or a cancer diagnosis than to reveal their eye color. This is great news for privacy researchers; it means that there is more work to do in this exciting field.

Privacy Laws Need Updating

A second policy shortcoming is that contemporary privacy laws frequently require organizations sharing data to make absolute assurances regarding the privacy of data subjects, but also allow sharing data without the consent of data subjects if data are "anonymized."

For example, Recital 26 of the European Union's General Data Protection Regulation (GDPR) exempts "anonymous information," which it defines as "information which does not relate to an identified or identifiable natural person or to personal data rendered anonymous in such a manner that the data subject is not or no longer identifiable."[11] The GDPR states that data are not anonymous if they allow the "singling out" of individuals. Meanwhile, Title 13, Section 9 of the US Code prohibits the US Census Bureau from making any publication "whereby the data furnished by any particular establishment or individual . . . can be identified."[12]

What these laws call for is mathematically impossible in the presence of auxiliary information. There is no way to assure that the absolute level of privacy protection required by these laws cannot be undone by linking against a dataset that doesn't exist today, but that is created and released next month.

In 2020, Aloni Cohen and Nissim modeled the GDPR's notion of singling out by creating a mathematically precise concept they call "predicate singling out."[13] They then showed the conditions under which predicate singling out is possible—that is, when a data release allows singling out as anticipated under the GDPR. Essentially, their article establishes a mathematical framework for understanding the mosaic effect. They also showed that while DP offers tunable protection against the mosaic effect, *k*-anonymity does not. That's a problem, as *k*-anonymity is widely used for anonymization under the GDPR.

Just as other laws have been updated in light of scientific progress, laws like the GDPR and Title 13 will need to be reinterpreted or rewritten to reflect what we now know about the nature of data as well as the mosaic effect.

3

FUTURE DIRECTIONS

So far I've presented differential privacy's state of the art. In this chapter, I'll present its likely future.

The Future Is Private

In January 1999, the chief executive officer of Sun Microsystems, at the time one of Silicon Valley's most valuable companies, said that concern about privacy was a "red herring. . . . You have zero privacy anyway," said Scott McNealy. "Get over it."[1]

What has become the most memorable quote of McNealy's career was shocking then and remains so now. Silicon Valley was in the middle of the dot-com boom, with its growing tech firms laying the groundwork for a new economy based on surveillance capitalism. Yet in Washington,

DC, companies like Sun were telling the US Congress not to pass laws protecting privacy. Instead, they promised that self-regulation would do a better job.

And what about the companies that didn't respect consumer privacy? Lawmakers need not worry about that remote possibility, the Silicon Valley leaders promised; if consumers felt that their personal data wasn't being respected, they could always vote with a mouse click and move to another provider. The free market would keep Silicon Valley in check.

Almost twenty-five years have passed, and interest in privacy has made a stunning comeback. This return is not driven by consumer flight; market consolidation has pretty much eliminated most consumers' ability to choose alternatives. Rather, lawmakers in Europe and California have started passing laws like the GDPR and California Consumer Privacy Act, respectively, and US regulators at agencies such as the Federal Trade Commission, Federal Communications Commission, and Consumer Financial Protection Bureau have begun insisting that companies respect the privacy concerns of their users.

Official Differential Privacy

The US Census Bureau was the first official statistics agency to adopt differential privacy; it will continue to do so, and it won't be the last. A growing number of government agencies are experimenting with differential privacy to protect official statistics, including the bureau's Post-Secondary Employment Outcomes experimental tabulations and the Department of Education's College Scorecard website.[2] I believe that officials at these agencies are motivated by the theoretical appeal of differential privacy, fact that DP allows for transparency in statistics making, and desire to be at the forefront of statistical methodology improvements. It isn't often that government agencies are leaders in technology, but with differential privacy, they can be.

Differential privacy also makes it possible for statistical officials to collaborate with a wide range of motivated, energetic, and frequently early career academics who care passionately about privacy. It lets them protect against the growing threat of auxiliary data and the mosaic effect. And perhaps most important, it gives officials mathematical dials that they can turn to adjust the privacy/utility trade-off.

Today we also see growing interest and even pilot projects in countries as different as the United Kingdom, Germany, Sweden, Japan, and Brazil.[3] Differential privacy is taking hold in government world-wide.

Aspirational Corporate Differential Privacy

Differential privacy may not be the privacy technology that most commercial privacy professionals want, but in the long run, it may be the only technology that they can use with safety and predictability. Companies that traffic

It isn't often that government agencies are leaders in technology, but with differential privacy, they can be.

in personal information need a technology that lets them exploit personal information while giving ironclad assurances that personal privacy will be protected. With DP, they can provide such guarantees and then tune the privacy budget so that they can still make a profit with personal data.

Several organizations that received kudos for their decision to adopt DP were later criticized for picking high values of epsilon, the PLB. For example, when researchers reverse engineered Apple Computer's DP implementations in 2017 and discovered that the company was using a value of 6 in MacOS and 14 in iOS, many were quick to criticize, including DP coinventor Frank McSherry. "Anything much bigger than 1 is not a very reassuring guarantee," McSherry told *WIRED*. "Using an epsilon value of 14 per day strikes me as relatively pointless."[4]

Likewise, when the US Census Bureau announced in 2021 that it would be using an epsilon setting of 19.61 for the 2020 Census redistricting product, one of the bureau's most vocal critics told the Associated Press that "the inventors of differential privacy regard such high epsilons as pointless."[5]

But having examined and discussed these high epsilon values with others, I don't think they are pointless or even really a problem—at least not in the short run. For companies like Apple and government agencies like the US Census Bureau to go through a data pipeline and make it differentially private is an incredibly time-consuming task. It requires completely rearchitecting the way that data are acquired, processed, transformed, and reported. Today much of this work can only be done by experts. When the system first turns on, the results are generally *terrible*, and the only way to improve the results is by cranking up the privacy loss. And for the commercial deployments at companies like Apple, Google, and Microsoft, DP is just one of many privacy protection technologies at work.

As the organizations gain more experience with differential privacy, however, they improve their systems. Sometimes this improvement is the result of better mechanisms and more attention to methodological details. Sometimes it comes from changing the queries and results that they report. At other times it comes from better mathematical proofs. Remember, early proofs regarding Google's differentially private stochastic gradient descent algorithm frequently had values of epsilon over 100, but by carefully reexamining the data pipeline, Google's scientists were able to show that the actual epsilon was much less—by a factor of 20 to 50 in some cases.[6]

Another lesson from the early years of DP is that not all epsilons are equal. A single question on a survey with an epsilon of 5 with differential privacy running locally would not offer much protection, in part because $e^{-5.0}$ is such a small number (a little less than 0.00674). But no reasonable organization would use DP in that manner. When the

Census Bureau announced that its effective epsilon was 19.61, it explained that the single PLB was being split up between thousands of individual queries. Although it is theoretically possible for an adversary to recombine all the individual queries and gain deep insight on the unprotected data, in practice such data machinations are far, far beyond today's state of the art—if they are even possible. Perhaps if the Census Bureau redoes its calculations some day in the future, it will discover that the effective epsilon for the 2020 Census was really something closer to 3 or 4.

When an organization adopts differential privacy, that organization needs to rigorously account for all uses of confidential data in its statistical productions. Without DP, such accounting rarely happens. Once an organization takes the first step, it can then work on lowering the PLB, which will typically involve redesigning aspects of its statistical products to be less disclosive. None of this would happen without the initial move to DP. For this reason, I think that people who focus on high values of epsilon are missing the bigger picture.

Combining DP with Trusted Execution Environments, Secure Multiparty Computation, and Homomorphic Encryption

Differential privacy is one of several privacy-enhancing technologies for statistical processing. Three other technologies are TEEs, MPC, and HE. It turns out that all of those technologies can also benefit from DP; in the future, we will likely see all three technologies being used together in hybrid systems.

TEEs are secure silicon data centers inside of some modern microprocessors; data can be encrypted so that it can only be decrypted by a specially approved code running inside the TEE. That code can perform a specific mathematical function, but it cannot execute arbitrary code, such as exporting the unencrypted confidential data. Intel's Software Guard Extensions (SGX) and Arm's TrustZone are examples of TEEs.

MPC and HE are two techniques for performing mathematical computations on encrypted data without the need to first decrypt. This is similar in spirit to local differential privacy, except that both MPC and HE will give exact, precise answers to computations of arbitrary complexity unless DP is explicitly added.

To return to our class example, with MPC the students could find the average without the teacher's help (as a trusted curator) and without sharing their grades with each other (a privacy-free scenario). Instead, each student would split their score in two unequal "shares" (for instance, the first student who scored 100 might split that number into 45 and 55), print each score on a poker chip, and throw all the chips into a bag. The students would then shake the bag, take out the twenty chips, and tally the scores; the result is the total scores of all the students.

Divide by ten and the result is the class average. With this simple protocol, nobody can trace a specific score to a specific student (assuming that the handwriting is indistinguishable).

As this example demonstrates, MPC can do a great job protecting data provenance; it makes it possible to perform arithmetic on data without knowing who provided which data elements. But it doesn't protect the output; the average score is still 98, so the student who got an 80 can still compute that the other nine students all got 100. That's why MPC (and HE) will most likely be deployed in conjunction with DP, which will add noise so that database reconstruction does not yield the true, confidential values.

Differential Privacy Grows Up

As DP matures and moves from the research labs to the world of applied statistics, DP practitioners will need to undergo the same sort of transformation that idealistic young undergraduates make when they graduate and enter the real world: they will need to start making compromises. Rather than letting the perfect be the enemy of the good, they will need to grab what wins they can in the ongoing project of improving (if not perfecting) the analysis of data that are sensitive or confidential.

Although we might wish to always use differential privacy when analyzing sensitive data, this will likely be an impossibility for many years into the future. In part this is because many kinds of data are not compatible with the DP analytic framework. These might be queries where the mathematical sensitivity (Δf) is infinity (∞). Or it might be a query where the idea of adding noise is nonsensical, like a keyword search run on a collection of college essays. In the future, we will likely develop new analytic frameworks that allow for private data analysis on a much broader range of data than is possible today. In the meantime, people need to get their jobs done.

New DP Variants

Differential privacy will continue to mature too. Just as (ε,δ)-DP offered practical improvements over ε-DP, and just as zCDP offered more improvements still, it's quite likely that privacy researchers will develop variants in the future that are easier to deploy and do a better job with real-world data. These variants will be properly seen as refinements of DP rather than a repudiation.

New DP Tools

What's most important for the continued success of differential privacy is a new generation of tools that make it easier for DP to be used by nonexperts. This includes tools for:

- Transforming and analyzing existing datasets with DP
- Automatically computing query sensitivity on existing pipelines
- Examining existing data pipelines and suggesting ways to recast those analyses in a manner that's compliant with DP
- Understanding what a high value of privacy loss means with respect to a specific data release
- Helping policymakers and other officials balance utility with privacy loss
- Helping policymakers and other officials decide when a single PLB can be shared between multiple analysts, and when it is OK to "reset" the budget with the arrival of new data

New Rules

Finally, the growing number of laws intended to protect privacy in the digital age need to be brought into line with our modern understanding of privacy. This is a tall order.

Lawmakers throughout the world are increasingly and rightfully concerned about the threats to privacy posed by data collection and use. As a result, many laws like the European Union's GDPR and the US Health Insurance Portability and Accountability Act of 1996 along with their implementation guidance codify specific approaches

What's needed is a new generation of laws that consider our modern understanding of privacy: that every data release may potentially have problems, and a balance is crucial between data utility and privacy.

regarding the use of personal information. Many of these approaches were based on lawmakers' aspirational hopes of how they wanted organizations to protect personal data rather than a formal analysis of what kinds of protection goals were mathematically possible.

It isn't that the lawmakers didn't believe the math; it's that the math didn't exist when the core concepts of today's privacy laws such as de-identification, anonymized data, and singling out were first articulated decades ago.

As lawmakers and other officials become knowledgeable about formal privacy approaches and differential privacy, they are caught in a conflict between knowing what is mathematically necessary to protect privacy and the legal requirements of their privacy laws. Frequently laws call for a level of protection that is mathematically nonsensical, while at the same time authorizing data processing practices that are mathematically inconsistent with legal objectives.

What's needed now is a new generation of laws that consider our modern understanding of privacy—an understanding that holds that every data release may potentially have lurking problems, and there must be a balance between data utility and privacy. In the future, laws will need to empower regulators to find that balance and then mandate it.

After all, there is no denying the math.

ACKNOWLEDGMENTS

This book dedicated to John Abowd and Christa Jones, because I never would have been in a position to write it without their support and trust.

I met Christa in 2015 shortly after I joined the US government's Privacy Research and Development (PrivacyRD) working group. Christa was participating in PrivacyRD to help the Census Bureau learn more about new developments in privacy-enhancing technologies. She introduced me to John, who had recently become the Census Bureau's chief scientist as well as associate director of research and methodology. They both invited me to move from NIST to the Census Bureau and help lead the adoption of differential privacy as well as modernize the bureau's Disclosure Review Board.

The Census Bureau is one of the most exciting places that I have ever worked. I was dropped into an organization that was mobilizing hundreds of thousands of people for the constitutionally mandated decennial count of population and housing, while at the same time continuing to produce more than a hundred other statistical products that support the economy of the United States. There I worked closely with Ron Jarmin, who performed the nonexclusive functions and duties of the Census Bureau's director from July 2017 through January 2019, when he became its deputy director and chief operating officer, and Phil Leclerc, a brilliant mathematician who helped lead the team that solved many of the mathematical challenges in developing and deploying the DP system for the 2020 Census, and spent many hours explaining various aspects of differential privacy to me. I also had many positive interactions with others within the Census Bureau during my time there. I especially wish to recognize Tammy Adams, Robert Ashmead, Steve Clark, Aref Dajani, John Eltinge, John Fattaleh, Nathan Goldschlag, Michael Hawes, Shelly Hedrick, Cynthia Hollingsworth, Beau Houser, Amy Lauger, Holly Newman, Krista Park, Michael Ratcliffe, Rolando Rodriguez, Teresa Sabol, William Sexton, Robert Sienkiewicz, Phyllis Singer, Lisa Stewart, Sara Sullivan, Ann Tozzi, and Victoria Velkoff. Moreover, I had the opportunity to work with several university faculty who had various government appointments, including Dan Kifer, Julia Lane, and Jerry Reiter.

Amy Brand at the MIT Press invited me to write this book in spring 2022. But *Differential Privacy* wouldn't have happened without the generous support of the Alfred P. Sloan Foundation, which gave me a grant to cover the cost of MIT's open-access fees and other costs associated with writing and publishing this book. My editor, Gita Manaktala, guided this project from inception through publication, and Matt Mahoney did a great job with the fact-checking and initial editing.

I am indebted to Ted Rall, who is both my illustrator and my friend of many decades. I hope that readers will agree with me that Ted's drawings bring a fun irreverence to what might otherwise be a dry, statistical topic. Ted is best known for his political cartoons and commentary, so my decision to use him as an illustrator caught some people by surprise. But make no mistake: the decision to use differential privacy and then deciding on the appropriate trade-off between statistical accuracy and privacy protection are very much political ones, similar to decisions such as teaching evolution to schoolchildren and counting every resident of the United States in the decennial census.

I wrote this book to give people who are not mathematicians, statisticians, or computer scientists a basic understanding of DP and its implications. This book does not teach you how to use differential privacy to analyze a confidential dataset, but it will give you an understanding of why you might want to do that as well as list of problems you are likely to encounter in the process.

Differential Privacy is based on my personal experiences in data privacy as a practitioner and journalist along with my job experiences as part of the Census Bureau's differential privacy team. As such, I approach this material as an advocate for the judicious use of differential privacy. I believe that we can broadly adopt many of DP's concepts to provide more meaningful privacy protections in today's data-rich world, even in cases where we cannot rigorously apply DP itself. I received answers to questions as well as useful comments on drafts of this book from Yasemin Acar (Paderborn University), Cynthia Dwork (Harvard University), Zeyu Ding (Binghamton University), Dan Kifer (Penn State), Julia Lane (NYU), Frank McSherry (Materialize), Joe Near (University of Maine), Kobbi Nissim (Georgetown University), M. Alejandra Parra-Orlandoni, Anand Sarwate (Rutgers University), Chinmay Shah (PyDP), Ted Selker (Selker Design Research), and Adam Smith (Boston University). Thanks to these and other reviewers—especially the anonymous ones arranged for by the MIT Press—this book is far better than what it would have been without their generous help.

The views expressed in this book are those of the author, and do not reflect the official policy or position of the Census Bureau, the Department of Commerce, or the US Government.

GLOSSARY

This glossary is based on the one in Simson Garfinkel, Joseph Near, Aref N. Dajani, Phyllis Singer, and Barbara Guttman, *De-Identifying Government Datasets: Techniques and Governance* (Gaithersburg, MD: National Institute of Standards and Technology, September 14, 2023), https://doi.org/10.6028/NIST.SP.800-188. Where noted, the definition is sourced from another publication.

Administrative data
Data collected or created by organizations (typically government agencies) as part of their administrative duties and not for statistical purposes, such as tax records and building permits.

Auxiliary information
Information available to a data hacker from other sources that can be used to undo privacy protection measures. Also known as *background knowledge*, *external knowledge*, or *side information*.

Confidentiality
Data confidentiality is a property of data, usually resulting from legislative measures, that prevents unauthorized disclosure. (OECD Glossary of Statistical Terms)

De-identification
A general term for any process of removing the association between a set of identifying data and the data subject. (ISO/TS 25237:2008[E] Health Informatics—Pseudonymization)

Differential privacy
A rigorous mathematical definition of disclosure that considers the risk that an individual's confidential data may be learned as a result of a mathematical analysis based on those data being made publicly available. (SP800–188)

Direct identifier
An attribute that alone enables the unique identification of a data principal (e.g., a person) within a specific operational context. (ISO/IEC 20889:2018)

Disclosure
Divulging of or provision of access to data. (ISO/TS 25237:2008)

Epsilon (ε)
Also called the *PLB*, epsilon is a parameter used in differential privacy that specifies the maximum amount of privacy loss that might result when data produced by a DP algorithm are made publicly available or otherwise used.

ε-differential privacy
The original or "pure" version of differential privacy, as defined in the 2006 paper.

(ε, δ)-differential privacy
Also known as approximate differential privacy, this form generally provides more accurate statistics for the same amount of privacy loss.

Federated learning
This approach makes it possible to train large-scale artificial intelligence and machine learning systems without first bringing all the data together in a single location. Privacy-preserving federated learning makes it possible to train with confidential data without revealing specifics about the confidential data.

Homomorphic encryption (HE)
A general-purpose approach for computing with data that are encrypted without needing to first decrypt the data. Differential privacy must still be added to the result of HE in order to limit the risk of privacy loss through database reconstruction.

Identifying information
Information that can be used to distinguish or trace an individual's identity, either alone or when combined with other information that is linked or linkable to a specific individual. (OMB M-17–12)

***k*-anonymity**
A data processing technique invented by Latanya Sweeney designed to make it harder to reidentify records in a dataset that works by having an expert remove the identifiers and then making sure that for every distinct combination of quasi-identifiers, there are at least *k* records with that combination.

Local model
An approach for using DP in which noise is added to each data element by the data subject before it is sent to the curator for statistics making, resulting in a curator that no longer needs to be trusted.

Microdata
The individual data records about people, establishments, or some other kind of "unit" that is being analyzed. Think of microdata as the rows of a spreadsheet, one per person.

Mosaic effect
The way that many small data releases or uses of confidential data about an individual can be combined to reveal information about that individual that was thought to be private, just as many small tiles in a mosaic combine to create a far more detailed picture.

Official statistics
Statistics produced by a government for some official purpose.

Parallel composition
When protecting statistics using differential privacy, statistics that enjoy parallel composition are drawn from disjoint datasets and thus do not result in additional privacy loss; the epsilons do not "add up."

Personal data
Any information relating to an identified or identifiable natural person (data subject). (ISO/TS 25237:2008)

Personally identifiable information
Information that can be used to distinguish or trace an individual's identity, either alone or when combined with other information that is linked or linkable to a specific individual. (A-130)

Postprocessing
Any statistical manipulation performed on data after the differential privacy mechanism is used.

Privacy
Freedom from intrusion into the private life or affairs of an individual when that intrusion results from the undue or illegal gathering and use of data about that individual. (ISO/IEC 2382–8:1998, definition 08-01-23)

Privacy loss
A measure of the extent to which a data release may reveal information that is specific to an individual. (SP800–188)

Privacy loss budget (PLB)
An upper bound on the cumulative total privacy loss for individuals. (SP800–188)

Reidentification
A process by which information is attributed to de-identified data in order to identify the individual to whom the de-identified data relate. (OECD-LEGAL-0433)

Rényi differential privacy
A refinement of (ε, δ)-differential privacy that offers an improved trade-off between privacy and utility when many statistics are released from a single confidential dataset.

Secure data enclaves
Places where people or programs can work with confidential data.

Secure multiparty computation (MPC)
An approach for working with sensitive data that uses sophisticated mathematical protocols based on cryptography that compute the desired statistic by sending messages between the different data owners. Differential privacy must be applied to the results of MPC in order to protect from database reconstruction.

Sequential composition
When computing multiple statistics from the same confidential data, the statistics are said to use "sequential" composition and the epsilons "add up."

Suppression
A common way for protecting sensitive cells in a table by preventing the data from appearing. Suppression is not differentially private.

Synthetic data
Data created using a variety of mathematical methods rather than simply being collected from observations.

Trusted curator
An entity that holds and protects confidential data, and provides it in a controlled environment for limited use, or uses the data to create statistics that can be released.

Trusted curator model
Also known as the central model of DP, this is the most common way in which DP is used: all the sensitive data are brought to one location, where the statistics are first made and then protected. (See local model.)

Zero-concentrated differential privacy (zCDP)
Another refinement of (ε, δ)-differential privacy that offers an improved tradeoff between privacy and utility when many statistics are released from a single confidential dataset.

NOTES

Preface

1. Simson Garfinkel, *De-Identification of Personal Information* (Gaithersburg, MD: National Institute of Standards and Technology, October 22, 2015), https://doi.org/10.6028/NIST.IR.8053.
2. "Court Case Tracker: Alabama v. U.S. Dep't of Commerce," Brennan Center for Justice, accessed March 2, 2024, https://www.brennancenter.org/our-work/court-cases/alabama-v-us-dept-commerce.
3. Paul M. Romer, "Mathiness in the Theory of Economic Growth," *American Economic Review* 105, no. 5 (May 2015): 89–93, https://doi.org/10.1257/aer.p20151066.
4. Daniel L. Oberski and Frauke Kreuter, "Differential Privacy and Social Science: An Urgent Puzzle," *Harvard Data Science Review* 2, no. 1 (January 31, 2020), https://doi.org/10.1162/99608f92.63a22079.
5. Simson Garfinkel, Joseph Near, Aref N. Dajani, Phyllis Singer, and Barbara Guttman, *De-Identifying Government Datasets: Techniques and Governance* (Gaithersburg, MD: National Institute of Standards and Technology, September 14, 2023), https://doi.org/10.6028/NIST.SP.800-188; Joseph P. Near and Xi He, "Differential Privacy for Databases," *Foundations and Trends® in Databases* 11, no. 2 (July 21, 2021): 109–225, https://doi.org/10.1561/1900000066.
6. Joseph P. Near and Chiké Abuah, *Programming Differential Privacy*, 2021, https://programming-dp.com; "Welcome," OpenDP, 2024, https://docs.opendp.org.

Introduction

1. Cynthia Dwork, Vitaly Feldman, Moritz Hardt, Toniann Pitassi, Omer Reingold, and Aaron Roth, "The Reusable Holdout: Preserving Validity in Adaptive Data Analysis," *Science* 349, no. 6248 (August 7, 2015): 636–638, https://doi.org/10.1126/science.aaa9375.
2. Cynthia Dwork, Frank McSherry, Kobbi Nissim, and Adam Smith, "Calibrating Noise to Sensitivity in Private Data Analysis," in *Theory of Cryptography*, ed. Shai Halevi and Tal Rabin (Berlin: Springer, 2006), 265–284, https://doi.org/10.1007/11681878_14.
3. "Pillar Investigates: USCCB Gen Sec Burrill Resigns after Sexual Misconduct Allegations," *Pillar*, July 20, 2021, https://www.pillarcatholic.com/p/pillar-investigates-usccb-gen-sec.

4. Michelle Boorstein and Heather Kelly, "Catholic Group Spent Millions on App Data That Tracked Gay Priests," *Washington Post*, May 2, 2023, https://www.washingtonpost.com/dc-md-va/2023/03/09/catholics-gay-priests-grindr-data-bishops.

5. Tore Dalenius, "Towards a Methodology for Statistical Disclosure Control," *Statistisk Tidskrift (Statistical Review)* 15 (1977): 429–444, https://hdl.handle.net/1813/111303.

6. Cynthia Dwork, "Differential Privacy," in *Automata, Languages, and Programming*, ed. Michele Bugliesi, Bart Preneel, Vladimiro Sassone, and Ingo Wegener (Berlin: Springer, 2006), 1–12, https://doi.org/10.1007/11787006_1.

7. Miles E. Smid, "Development of the Advanced Encryption Standard," *Journal of Research of the National Institute of Standards and Technology* 126, no. 126024 (2021), https://doi.org/10.6028/jres.126.024.

8. Lance J. Hoffman and William F. Miller, "Getting a Personal Dossier from a Statistical Data Bank," *Datamation* 16, no. 5 (1970): 74–75.

9. Dalenius, "Towards a Methodology for Statistical Disclosure Control."

10. Irit Dinur and Kobbi Nissim, "Revealing Information while Preserving Privacy," in *Proceedings of the Twenty-Second ACM SIGMOD-SIGACT-SIGART Symposium on Principles of Database Systems* (New York: Association for Computing Machinery, 2003), 202–210, https://doi.org/10.1145/773153.773173.

11. Dinur and Nissim, "Revealing Information while Preserving Privacy," 208.

12. Cynthia Dwork and Kobbi Nissim, "Privacy-Preserving Datamining on Vertically Partitioned Databases," Microsoft, 2004, 528–544, https://www.microsoft.com/en-us/research/publication/privacy-preserving-datamining-on-vertically-partitioned-databases.

13. Avrim Blum, Cynthia Dwork, Frank McSherry, and Kobbi Nissim, "Practical Privacy: The SuLQ Framework," in *Proceedings of the Twenty-Fourth ACM SIGMOD-SIGACT-SIGART Symposium on Principles of Database Systems* (New York: Association for Computing Machinery, 2005), 128–138, https://doi.org/10.1145/1065167.1065184.

14. Cynthia Dwork, "Reminiscences," *Journal of Privacy and Confidentiality* 8, no. 1 (December 2, 2018), https://doi.org/10.29012/jpc.702.

15. Dwork, McSherry, Nissim, and Smith, "Calibrating Noise to Sensitivity in Private Data Analysis."

16. Dwork, "Differential Privacy."

17. Cynthia Dwork and Moni Naor, "On the Difficulties of Disclosure Prevention in Statistical Databases or the Case for Differential Privacy," *Journal of Privacy and Confidentiality* 2, no. 1 (September 1, 2010), https://doi.org/10.29012/jpc.v2i1.585.

18. Ashwin Machanavajjhala, Daniel Kifer, John Abowd, Johannes Gehrke, and Lars Vilhuber, "Privacy: Theory Meets Practice on the Map," in *2008 IEEE 24th International Conference on Data Engineering* (New York: IEEE, 2008), 277–286, https://doi.org/10.1109/ICDE.2008.4497436.
19. Úlfar Erlingsson, Vasyl Pihur, and Aleksandra Korolova, "RAPPOR: Randomized Aggregatable Privacy-Preserving Ordinal Response," in *Proceedings of the 2014 ACM SIGSAC Conference on Computer and Communications Security* (New York: Association for Computing Machinery, 2014), 1054–1067, https://doi.org/10.1145/2660267.2660348.
20. Differential Privacy Team, "Learning with Privacy at Scale," Apple Machine Learning Research, December 2017, https://machinelearning.apple.com/research/learning-with-privacy-at-scale; Bolin Ding, Janardhan Kulkarni, and Sergey Yekhanin, "Collecting Telemetry Data Privately," in *Proceedings of the 31st International Conference on Neural Information Processing Systems* (Red Hook, NY: Curran Associates, Inc., 2017), 3574–3583, https://dl.acm.org/doi/10.5555/3294996.3295115; Katie Tezapsidis, "Uber Releases Open Source Project for Differential Privacy," *Uber Privacy and Security* (blog), July 13, 2017, https://medium.com/uber-security-privacy/differential-privacy-open-source-7892c82c42b6.
21. Simson L. Garfinkel, "Modernizing Disclosure Avoidance: Report on the 2020 Disclosure Avoidance Subsystem as Implemented for the 2018 End-to-End Test" (paper presented at the 2017 Census Scientific Advisory Committee, Suitland, MD, September 15, 2017), https://www2.census.gov/cac/sac/meetings/2017-09/garfinkel-modernizing-disclosure-avoidance.pdf.
22. John M. Abowd, "The U.S. Census Bureau Adopts Differential Privacy," in *Proceedings of the 24th ACM SIGKDD International Conference on Knowledge Discovery and Data Mining* (New York: Association for Computing Machinery, 2018), 2867, https://doi.org/10.1145/3219819.3226070; *2020 Census Operational Plan: A New Design for the 21st Century* (Suitland, MD: US Census Bureau, December 2018), https://www2.census.gov/programs-surveys/decennial/2020/program-management/planning-docs/2020-oper-plan4.pdf.
23. "First United States Census, 1790," Mount Vernon, accessed April 14, 2024, https://www.mountvernon.org/library/digitalhistory/digital-encyclopedia/article/first-united-states-census-1790.
24. US Census Bureau, "Authorizing the First Census—The Significance of Population Data," United States Census Bureau, accessed September 9, 2023, https://www.census.gov/programs-surveys/sis/activities/history/first-census.html.

25. "1790 Census Records," National Archives, October 28, 2020, https://www.archives.gov/research/census/1790.
26. US Census Bureau, "Monograph: Census Confidentiality and Privacy: 1790–2002," United States Census Bureau, accessed September 10, 2023, https://www.census.gov/library/publications/2003/comm/monograph-confidentiality-privacy.html.
27. Carroll D. Wright and William C. Hunt, *History and Growth of the United States Census: 1790–1890* (Washington, DC: Government Printing Office, 1900).
28. Samuel D. Warren and Louis D. Brandeis, "The Right to Privacy," *Harvard Law Review* 4, no. 5 (1890): 193–220, https://doi.org/10.2307/1321160.
29. Dan Bouk, *Democracy's Data* (New York: Macmillan, 2022), 27, https://us.macmillan.com/books/9780374602543/democracysdata.
30. "1940 Census Records," National Archives, September 8, 2016, https://www.archives.gov/research/census/1940.
31. Bouk, *Democracy's Data,* 215.
32. J. R. Minkel, "Confirmed: The U.S. Census Bureau Gave Up Names of Japanese-Americans in WW II," *Scientific American*, March 30, 2007, https://www.scientificamerican.com/article/confirmed-the-us-census-b.
33. United States Bureau of the Census, *U.S. Census of Population: 1960: Availability of Published and Unpublished Data* (Washington, DC: Department of Commerce, 1964).
34. United States Bureau of the Census, *U.S. Census of Population*.
35. Fertility and Family Statistics Branch, "Frequently Asked Questions about Same-Sex Couple Households," US Census Bureau, August 2013, https://www2.census.gov/topics/families/same-sex-couples/faq/sscplfactsheet-final.pdf.

Chapter 1

1. H. D. Arnold and Lloyd Esenschied, "Transatlantic Radio Telephony," *Bell Systems Technical Journal* 2 (October 1923): 116–144, http://archive.org/details/bstj2-4-116.
2. Daniel Barth-Jones, "The 'Re-Identification' of Governor William Weld's Medical Information: A Critical Re-Examination of Health Data Identification Risks and Privacy Protections, Then and Now," SSRN Scholarly Paper, Rochester, NY, July 1, 2012, https://doi.org/10.2139/ssrn.2076397.
3. Office for Civil Rights, "Guidance regarding Methods for De-Identification of Protected Health Information in Accordance with the Health Insurance Portability and Accountability Act (HIPAA) Privacy Rule," US Department of

Health and Human Services, October 25, 2022, https://www.hhs.gov/hipaa/for-professionals/privacy/special-topics/de-identification/index.html.
4. Geeta Anand, "A Spirited Weld Goes Home," *Boston Globe*, May 20, 1996, https://www.newspapers.com/image/440942660.
5. Latanya Sweeney, "Only You, Your Doctor, and Many Others May Know," *Technology Science*, accessed October 21, 2023, https://techscience.org/a/2015092903.
6. "Baser v. VA, No. 13–12591, 2014 WL 4897290 (E.D. Mich., Sept. 30, 2014) (Hood, J.)," Office of Information Policy, US Department of Justice, September 30, 2014, https://www.justice.gov/oip/baser-v-va-no-13-12591-2014-wl-4897290-ed-mich-sept-30-2014-hood-j.
7. William Howard Taft, "By the President of the United States of America: A Proclamation," March 15, 1910, https://www.census.gov/history/img/proclamation1910-artifact.jpg.
8. Shelly Hedrick, "Census Protections Evolve Continuously to Address Emerging Threats," United States Census Bureau, February 3, 2020, https://www.census.gov/library/stories/2020/02/through-the-decades-how-the-census-bureau-protects-your-privacy.html.
9. Federal Committee on Statistical Methodology, *Statistical Policy Working Paper 22: Report on the Statistical Disclosure Limitation Methodology* (Washington, DC: Office of Management and Budget, December 2005), https://www.hhs.gov/sites/default/files/spwp22.pdf.
10. Tore Dalenius, "Towards a Methodology for Statistical Disclosure Control," *Statistisk Tidskrift* (*Statistical Review*) 15 (1977): 429–444, https://hdl.handle.net/1813/111303.
11. Cynthia Dwork, "Differential Privacy," in *Automata, Languages, and Programming*, ed. Michele Bugliesi, Bart Preneel, Vladimiro Sassone, and Ingo Wegener (Berlin: Springer, 2006), 1–12, https://doi.org/10.1007/11787006_1.
12. Jenny Ashcraft, "Destruction of the 1890 Census," *Fishwrap* (blog), October 29, 2023, https://blog.newspapers.com/destruction-of-the-1890-census.
13. Edouard Mathieu, Hannah Ritchie, Lucas Rodés-Guirao, Cameron Appel, Charlie Giattino, Joe Haskell, Bobbie Macdonald, et al., "COVID-19: Google Mobility Trends," Our World in Data, March 5, 2020, https://ourworldindata.org/covid-google-mobility-trends.
14. Ahmet Aktay, Shailesh Bavadekar, Gwen Cossoul, John Davis, Damien Desfontaines, Alex Fabrikant, Evgeniy Gabrilovich, et al., "Google COVID-19 Community Mobility Reports: Anonymization Process Description (Version 1.1)," arXiv, November 3, 2020, https://doi.org/10.48550/arXiv.2004.04145.

15. Rebecca Kraus, "Statistical Déjà Vu: The National Data Center Proposal of 1965 and Its Descendants," *Journal of Privacy and Confidentiality* 5, no. 1 (August 1, 2013), https://doi.org/10.29012/jpc.v5i1.624.
16. Cynthia Dwork, Frank McSherry, Kobbi Nissim, and Adam Smith, "Calibrating Noise to Sensitivity in Private Data Analysis," in *Theory of Cryptography*, ed. Shai Halevi and Tal Rabin (Berlin: Springer, 2006), 265–284, https://doi.org/10.1007/11681878_14.
17. Daniel Alabi, Audra McMillan, Jayshree Sarathy, Adam Smith, and Salil Vadhan, "Differentially Private Simple Linear Regression," *Proceedings on 23rd Privacy Enhancing Technologies Symposium (PoPETS '22)* 2 (2022): 184–204, https://salil.seas.harvard.edu/publications/differentially-private-simple-linear-regression-0.
18. Caroline Medina and Lindsay Mahowald, "Key Issues Facing People with Intersex Traits," Center for American Progress, October 26, 2021, https://www.americanprogress.org/article/key-issues-facing-people-intersex-traits.
19. Cynthia Dwork and Guy N. Rothblum, "Concentrated Differential Privacy," arXiv, March 16, 2016, https://doi.org/10.48550/arXiv.1603.01887.
20. Mark Bun and Thomas Steinke, "Concentrated Differential Privacy: Simplifications, Extensions, and Lower Bounds," in *Proceedings, Part I, of the 14th International Conference on Theory of Cryptography—Volume 9985* (Berlin: Springer, 2016), 635–658, https://doi.org/10.1007/978-3-662-53641-4_24.
21. National Cancer Institute Press Office, "Three Breast Cancer Gene Alterations in Jewish Community," US Department of Health and Human Services, May 20, 1997.
22. Martin Abadi, Andy Chu, Ian Goodfellow, H. Brendan McMahan, Ilya Mironov, Kunal Talwar, and Li Zhang, "Deep Learning with Differential Privacy," in *Proceedings of the 2016 ACM SIGSAC Conference on Computer and Communications Security* (New York: Association for Computing Machinery, 2016), 308–318, https://doi.org/10.1145/2976749.2978318.
23. Milad Nasr, Jamie Hayes, Thomas Steinke, Borja Balle, Florian Tramèr, Matthew Jagielski, Nicholas Carlini, and Andreas Terzis, "Tight Auditing of Differentially Private Machine Learning," 32nd USENIX Security Symposium, 2023, 1631–1648, https://www.usenix.org/conference/usenixsecurity23/presentation/nasr.
24. Rolando Rodríguez, "Confidentiality for American Community Survey Respondents" (paper present at the 2023 ACS Data Users Conference, May 16–18, 2023), https://acsdatacommunity.prb.org/p/conferences.
25. Arthur Kennickell and Julia Lane, "Measuring the Impact of Data Protection Techniques on Data Utility: Evidence from the Survey of Consumer

Finances," in *Privacy in Statistical Databases*, ed. Josep Domingo-Ferrer and Luisa Franconi (Berlin: Springer, 2006), 4302: 291–303, https://doi.org/10.1007/11930242_25.
26. Stanley L. Warner, "Randomized Response: A Survey Technique for Eliminating Evasive Answer Bias," *Journal of the American Statistical Association*, March 1, 1965, https://www.tandfonline.com/doi/abs/10.1080/01621459.1965.10480775.
27. Úlfar Erlingsson, Vasyl Pihur, and Aleksandra Korolova, "RAPPOR: Randomized Aggregatable Privacy-Preserving Ordinal Response," in *Proceedings of the 2014 ACM SIGSAC Conference on Computer and Communications Security* (New York: Association for Computing Machinery, 2014), 1054–1067, https://doi.org/10.1145/2660267.2660348.
28. Dave Buckley, Mark Durkee, Joe Near, and David Darais, "Privacy Attacks in Federated Learning," *Responsible Technology Adoption Unit* (blog), January 24, 2024, https://rtau.blog.gov.uk/2024/01/24/privacy-attacks-in-federated-learning.
29. Shuang Song, Kamalika Chaudhuri, and Anand D. Sarwate, "Stochastic Gradient Descent with Differentially Private Updates," in *2013 IEEE Global Conference on Signal and Information Processing* (New York: IEEE, 2013), 245–248, https://doi.org/10.1109/GlobalSIP.2013.6736861.
30. Minxin Du, Xiang Yue, Sherman S. M. Chow, Tianhao Wang, Chenyu Huang, and Huan Sun, "DP-Forward: Fine-Tuning and Inference on Language Models with Differential Privacy in Forward Pass," in *Proceedings of the 2023 ACM SIGSAC Conference on Computer and Communications Security* (New York: Association for Computing Machinery, 2023), 2665–2679, https://doi.org/10.1145/3576915.3616592.
31. Dalenius, "Towards a Methodology for Statistical Disclosure Control."
32. I. P. Fellegi, "On the Question of Statistical Confidentiality," *Journal of the American Statistical Association* 67, no. 337 (March 1, 1972): p. 8, https://doi.org/10.1080/01621459.1972.10481199.
33. Irit Dinur and Kobbi Nissim, "Revealing Information while Preserving Privacy," in *Proceedings of the Twenty-Second ACM SIGMOD-SIGACT-SIGART Symposium on Principles of Database Systems* (New York: Association for Computing Machinery, 2003), 202–210, https://doi.org/10.1145/773153.773173.
34. Cynthia Dwork, "Reminiscences," *Journal of Privacy and Confidentiality* 8, no. 1 (December 2, 2018), https://doi.org/10.29012/jpc.702.
35. Cynthia Dwork and Jonathan Ullman, "The Fienberg Problem: How to Allow Human Interactive Data Analysis in the Age of Differential Privacy,"

Journal of Privacy and Confidentiality 8, no. 1 (December 21, 2018), https://doi.org/10.29012/jpc.687.

36. Frank D. McSherry, "Privacy Integrated Queries: An Extensible Platform for Privacy-Preserving Data Analysis," in *Proceedings of the 2009 ACM SIGMOD International Conference on Management of Data* (New York: Association for Computing Machinery, 2009), 19–30, https://doi.org/10.1145/1559845.1559850.

37. Frank McSherry, "Differential Privacy for Dummies," *GitHub* (blog), February 3, 2016, https://github.com/frankmcsherry/blog/blob/master/posts/2016-02-03.md; Frank McSherry, "Response to Criticism of 'Fools Gold' Article," *GitHub* (blog), May 19, 2016, https://github.com/frankmcsherry/blog/blob/master/posts/2016-05-19.md.

38. Frank McSherry and Kunal Talwar, "Mechanism Design via Differential Privacy," in *48th Annual IEEE Symposium on Foundations of Computer Science (FOCS'07)* (New York: IEEE, 2007), 94–103, https://doi.org/10.1109/FOCS.2007.66.

39. Yevgeniy Dodis, Leonid Reyzin, and Adam Smith, "Fuzzy Extractors: How to Generate Strong Keys from Biometrics and Other Noisy Data," in *Advances in Cryptology—EUROCRYPT 2004*, ed. Christian Cachin, Jan L. Camenisch, and Rafail Ostrovsky (Berlin: Springer, 2004), 523–540, https://doi.org/10.1007/978-3-540-24676-3_31.

40. Ilya Mironov, "On Significance of the Least Significant Bits for Differential Privacy," in *Proceedings of the 2012 ACM Conference on Computer and Communications Security* (New York: Association for Computing Machinery, 2012), 650, https://doi.org/10.1145/2382196.2382264.

41. Ilya Mironov, "Rényi Differential Privacy," in *30th Computer Security Foundations Symposium* (New York: IEEE, 2017), 263–275, https://doi.org/10.1109/CSF.2017.11; Úlfar Erlingsson, Vitaly Feldman, Ilya Mironov, Ananth Raghunathan, Kunal Talwar, and Abhradeep Thakurta, "Amplification by Shuffling: From Local to Central Differential Privacy via Anonymity," in *Proceedings of the Thirtieth Annual ACM-SIAM Symposium on Discrete Algorithms* (Philadelphia: Society for Industrial and Applied Mathematics, 2019), 2468–2479.

42. McSherry and Talwar, "Mechanism Design via Differential Privacy"; Erlingsson et al., "Amplification by Shuffling."

43. Cynthia Dwork and Aaron Roth, *The Algorithmic Foundations of Differential Privacy* (Boston: Now Publishers, 2013), 211–407, https://doi.org/10.1561/0400000042.

44. Arvin Narayanan and Vitaly Shmatikov, "Robust De-Anonymization of Large Sparse Datasets," in *2008 IEEE Symposium on Security and Privacy* (New York: IEEE, 2008), https://ieeexplore.ieee.org/document/4531148.

45. Reza Shokri, Marco Stronati, Congzheng Song, and Vitaly Shmatikov, "Membership Inference Attacks against Machine Learning Models," in *2017 IEEE Symposium on Security and Privacy* (New York: IEEE, 2017), 3–18, https://doi.org/10.1109/SP.2017.41.

46. Eugene Bagdasaryan, Omid Poursaeed, and Vitaly Shmatikov, "Differential Privacy Has Disparate Impact on Model Accuracy," in *Advances in Neural Information Processing Systems*, vol. 32 (Red Hook, NY: Curran Associates, Inc., 2019), https://proceedings.neurips.cc/paper_files/paper/2019/hash/fc0de4e0396fff257ea362983c2dda5a-Abstract.html.

47. Ashwin Machanavajjhala, Daniel Kifer, John Abowd, Johannes Gehrke, and Lars Vilhuber, "Privacy: Theory Meets Practice on the Map," in *2008 IEEE 24th International Conference on Data Engineering* (New York: IEEE, 2008), 277–286, https://doi.org/10.1109/ICDE.2008.4497436.

48. Úlfar Erlingsson, Vasyl Pihur, and Aleksandra Korolova, "RAPPOR: Randomized Aggregatable Privacy-Preserving Ordinal Response," in *Proceedings of the 2014 ACM SIGSAC Conference on Computer and Communications Security* (New York: Association for Computing Machinery, 2014), 1054–1067, https://doi.org/10.1145/2660267.2660348.

49. Erlingsson et al., "Amplification by Shuffling."

50. Differential Privacy Team, "Learning with Privacy at Scale," Apple Machine Learning Research, December 2017, https://machinelearning.apple.com/research/learning-with-privacy-at-scale.

51. Bolin Ding, Janardhan Kulkarni, and Sergey Yekhanin, "Collecting Telemetry Data Privately," in *Proceedings of the 31st International Conference on Neural Information Processing Systems* (Red Hook, NY: Curran Associates, Inc., 2017), 3574–3583, https://dl.acm.org/doi/10.5555/3294996.3295115.

52. Kashmir Hill, "'God View': Uber Allegedly Stalked Users for Party-Goers' Viewing Pleasure (Updated)," *Forbes*, October 3, 2014, https://www.forbes.com/sites/kashmirhill/2014/10/03/god-view-uber-allegedly-stalked-users-for-party-goers-viewing-pleasure.

53. Chris Welch, "Uber Will Pay $20,000 Fine in Settlement over 'God View' Tracking," *Verge*, January 6, 2016, https://www.theverge.com/2016/1/6/10726004/uber-god-mode-settlement-fine; "Uber Settles FTC Allegations That It Made Deceptive Privacy and Data Security Claims," Federal Trade Commission," August 15, 2017, https://www.ftc.gov/news-events/news/press-releases/2017/08/uber-settles-ftc-allegations-it-made-deceptive-privacy-data-security-claims.

54. Katie Tezapsidis, "Uber Releases Open Source Project for Differential Privacy," *Uber Privacy and Security* (blog), July 13, 2017, https://medium.com/uber-security-privacy/differential-privacy-open-source-7892c82c42b6.

Chapter 2

1. W. Diffie and M. Hellman, "New Directions in Cryptography," *IEEE Transactions on Information Theory* 22, no. 6 (November 1976): 644–654, https://doi.org/10.1109/TIT.1976.1055638; Ronald L. Rivest, Adi Shamir, and Len Adleman, "A Method for Obtaining Digital Signatures and Public-Key Cryptosystems," DSpace@MIT, April 1977, https://dspace.mit.edu/handle/1721.1/148910; Loren M. Kohnfelder, "Towards a Practical Public-Key Cryptosystem" (master's thesis, Massachusetts Institute of Technology, 1978), https://dspace.mit.edu/handle/1721.1/15993; Klint Finley, "It's Time to Encrypt the Entire Internet," *WIRED*, April 17, 2014, https://www.wired.com/2014/04/https.

2. Cynthia Dwork, Vitaly Feldman, Moritz Hardt, Toniann Pitassi, Omer Reingold, and Aaron Roth, "The Reusable Holdout: Preserving Validity in Adaptive Data Analysis," *Science* 349, no. 6248 (August 7, 2015): 636–638, https://doi.org/10.1126/science.aaa9375.

3. Cynthia Dwork, "Differential Privacy," in *Automata, Languages, and Programming*, ed. Michele Bugliesi, Bart Preneel, Vladimiro Sassone, and Ingo Wegener (Berlin: Springer, 2006), 1–12, https://doi.org/10.1007/11787006_1.

4. Shiva P. Kasiviswanathan and Adam Smith, "On the 'Semantics' of Differential Privacy: A Bayesian Formulation," *Journal of Privacy and Confidentiality* 6, no. 1 (June 1, 2014), https://doi.org/10.29012/jpc.v6i1.634.

5. Daniel Kifer, John M. Abowd, Robert Ashmead, Ryan Cumings-Menon, Philip Leclerc, Ashwin Machanavajjhala, William Sexton, and Pavel Zhuravlev, "Bayesian and Frequentist Semantics for Common Variations of Differential Privacy: Applications to the 2020 Census," arXiv, September 7, 2022, https://doi.org/10.48550/arXiv.2209.03310.

6. Matthew Fredrikson, Eric Lantz, and Somesh Jha, Simon Lin, David Page, and Thomas Ristenpart, "Privacy in Pharmacogenetics: An End-to-End Case Study of Personalized Warfarin Dosing," 23rd USENIX Security Symposium, August 20–22, 2014, 17–32, https://www.usenix.org/conference/usenixsecurity14/technical-sessions/presentation/fredrikson_matthew.

7. US Census Bureau, *Disclosure Avoidance for the 2020 Census: An Introduction* (Washington, DC: US Government Printing Office, November 2021), https://www2.census.gov/library/publications/decennial/2020/2020-census-disclosure-avoidance-handbook.pdf.

8. Cynthia Dwork, Moni Naor, Toniann Pitassi, and Guy N. Rothblum, "Differential Privacy under Continual Observation," in *Proceedings of the Forty-Second ACM Symposium on Theory of Computing* (New York: Association for Computing Machinery, 2010), 715–724, https://doi.org/10.1145/1806689.1806787;

Palak Jain, Sofya Raskhodnikova, Satchit Sivakumar, and Adam Smith, "The Price of Differential Privacy under Continual Observation," in *Proceedings of the 40th International Conference on Machine Learning* (Honolulu: JMLR, 2023), 202:14654–14678.
9. Arthur Miller, "Personal Privacy in the Computer Age: The Challenge of a New Technology in an Information-Oriented Society," *Michigan Law Review* 67, no. 6 (April 1, 1969): 1089–1247, https://repository.law.umich.edu/mlr/vol67/iss6/2.
10. Stephen Shankland, "Google Begins Blurring Faces in Street View," *CNET*, May 13, 2008, https://www.cnet.com/culture/google-begins-blurring-faces-in-street-view.
11. "Recital 26: Not Applicable to Anonymous Data," GDPR, April 27, 2016, https://gdpr.eu/recital-26-not-applicable-to-anonymous-data.
12. "Information as Confidential; Exception," Title 13—Census, August 31, 1954, https://www.govinfo.gov/content/pkg/USCODE-2022-title13/pdf/USCODE-2022-title13-chap1-subchapI-sec9.pdf.
13. Aloni Cohen and Kobbi Nissim, "Towards Formalizing the GDPR's Notion of Singling Out," *Proceedings of the National Academy of Sciences* 117, no. 15 (April 14, 2020): 8344–8352, https://doi.org/10.1073/pnas.1914598117.

Chapter 3

1. Polly Sprenger, "Sun on Privacy: 'Get Over It,'" *WIRED*, January 26, 1999, https://www.wired.com/1999/01/sun-on-privacy-get-over-it.
2. Andrew David Foote, Ashwin Machanavajjhala, and Kevin McKinney, "Releasing Earnings Distributions Using Differential Privacy: Disclosure Avoidance System for Post-Secondary Employment Outcomes (PSEO)," *Journal of Privacy and Confidentiality* 9, no. 2 (October 18, 2019), https://doi.org/10.29012/jpc.722; US Department of Education, *Technical Documents: College Scorecard Data by Field of Study*, February 2024, https://collegescorecard.ed.gov/assets/FieldOfStudyDataDocumentation.pdf.
3. Iain Dove, "Applying Differential Privacy Protection to ONS Mortality Data, Pilot Study," Office for National Statistics, August 20, 2021, https://www.ons.gov.uk/peoplepopulationandcommunity/birthsdeathsandmarriages/deaths/methodologies/applyingdifferentialprivacyprotectiontoonsmortalitydatapilotstudy; Patrick Kühtreiber, Viktoriya Pak, and Delphine Reinhardt, "Replication: The Effect of Differential Privacy Communication on German Users' Comprehension and Data Sharing Attitudes," in *Proceedings of the Eighteenth USENIX Conference on Usable Privacy and Security* (Berkeley, CA: USENIX Association, 2022), 117–134; Hamid Ebadi, David Sands, and Gerardo Schneider,

"Differential Privacy: Now It's Getting Personal," *ACM SIGPLAN Notices* 50, no. 1 (January 14, 2015): 69–81, https://doi.org/10.1145/2775051.2677005; Shinsuke Ito, Takayuki Miura, Hiroto Akatsuka, and Masayuki Terada, "Differential Privacy and Its Applicability for Official Statistics in Japan—A Comparative Study Using Small Area Data from the Japanese Population Census," in *Privacy in Statistical Databases: UNESCO Chair in Data Privacy, International Conference, PSD 2020, Tarragona, Spain, September 23–25, 2020, Proceedings* (Berlin: Springer, 2020), 337–352, https://doi.org/10.1007/978-3-030-57521-2_24; Gabriel H. Nunes, Mário S. Alvim, and Annabelle McIver, "A Formal Quantitative Study of Privacy in the Publication of Official Educational Censuses in Brazil," in *Anais Do Concurso de Teses e Dissertações (CTD)* (Anais do XXXV Concurso de Teses e Dissertações, SBC, 2022), 61–70, https://doi.org/10.5753/ctd.2022.223158.

4. Andy Greenberg, "How One of Apple's Key Privacy Safeguards Falls Short," *WIRED*, September 15, 2017, https://www.wired.com/story/apple-differential-privacy-shortcomings.

5. Mike Schneider, "Census Releases Guidelines for Controversial Privacy Tool," AP News, June 9, 2021, https://apnews.com/article/business-census-2020-55519b7534bd8d61028020d79854e909.

6. Martin Abadi, Andy Chu, Ian Goodfellow, H. Brendan McMahan, Ilya Mironov, Kunal Talwar, and Li Zhang, "Deep Learning with Differential Privacy," in *Proceedings of the 2016 ACM SIGSAC Conference on Computer and Communications Security* (New York: Association for Computing Machinery, 2016), 308–318, https://doi.org/10.1145/2976749.2978318; Milad Nasr, Jamie Hayes, Thomas Steinke, Borja Balle, Florian Tramèr, Matthew Jagielski, Nicholas Carlini, and Andreas Terzis, "Tight Auditing of Differentially Private Machine Learning," 32nd USENIX Security Symposium, 2023, 1631–1648, https://www.usenix.org/conference/usenixsecurity23/presentation/nasr.

BIBLIOGRAPHY

Abadi, Martin, Andy Chu, Ian Goodfellow, H. Brendan McMahan, Ilya Mironov, Kunal Talwar, and Li Zhang. "Deep Learning with Differential Privacy." In *Proceedings of the 2016 ACM SIGSAC Conference on Computer and Communications Security*, 308–318. New York: Association for Computing Machinery, 2016. https://doi.org/10.1145/2976749.2978318.

Abowd, John M. "The U.S. Census Bureau Adopts Differential Privacy." In *Proceedings of the 24th ACM SIGKDD International Conference on Knowledge Discovery and Data Mining*, 2867. New York: Association for Computing Machinery, 2018. https://doi.org/10.1145/3219819.3226070.

Aktay, Ahmet, Shailesh Bavadekar, Gwen Cossoul, John Davis, Damien Desfontaines, Alex Fabrikant, Evgeniy Gabrilovich, et al. "Google COVID-19 Community Mobility Reports: Anonymization Process Description (Version 1.1)." arXiv, November 3, 2020. https://doi.org/10.48550/arXiv.2004.04145.

Alabi, Daniel, Audra McMillan, Jayshree Sarathy, Adam Smith, and Salil Vadhan. "Differentially Private Simple Linear Regression." *Proceedings on 23rd Privacy Enhancing Technologies Symposium (PoPETS '22)* 2 (2022): 184–204. https://salil.seas.harvard.edu/publications/differentially-private-simple-linear-regression-0.

Anand, Geeta. "A Spirited Weld Goes Home." *Boston Globe*, May 20, 1996. https://www.newspapers.com/image/440942660.

Arnold, H. D., and Lloyd Esenschied. "Transatlantic Radio Telephony." *Bell Systems Technical Journal* 2 (October 1923): 116–144. http://archive.org/details/bstj2-4-116.

Ashcraft, Jenny. "Destruction of the 1890 Census." *Fishwrap* (blog), October 29, 2023. https://blog.newspapers.com/destruction-of-the-1890-census.

Bagdasaryan, Eugene, Omid Poursaeed, and Vitaly Shmatikov. "Differential Privacy Has Disparate Impact on Model Accuracy." In *Advances in Neural Information Processing Systems*. Vol. 32. Red Hook, NY: Curran Associates, Inc., 2019. https://proceedings.neurips.cc/paper_files/paper/2019/hash/fc0de4e0396fff257ea362983c2dda5a-Abstract.html. Barth-Jones, Daniel. "The 'Re-

Identification' of Governor William Weld's Medical Information: A Critical Re-Examination of Health Data Identification Risks and Privacy Protections, Then and Now." SSRN Scholarly Paper. Rochester, NY, July 1, 2012. https://doi.org/10.2139/ssrn.2076397.

"Baser v. VA, No. 13–12591, 2014 WL 4897290 (E.D. Mich., Sept. 30, 2014) (Hood, J.)." Office of Information Policy, US Department of Justice, September 30, 2014. https://www.justice.gov/oip/baser-v-va-no-13-12591-2014-wl-4897290-ed-mich-sept-30-2014-hood-j.

Blum, Avrim, Cynthia Dwork, Frank McSherry, and Kobbi Nissim. "Practical Privacy: The SuLQ Framework." In *Proceedings of the Twenty-Fourth ACM SIGMOD-SIGACT-SIGART Symposium on Principles of Database Systems*, 128–138. New York: Association for Computing Machinery, 2005. https://doi.org/10.1145/1065167.1065184.

Boorstein, Michelle, and Heather Kelly. "Catholic Group Spent Millions on App Data That Tracked Gay Priests." *Washington Post*, May 2, 2023. https://www.washingtonpost.com/dc-md-va/2023/03/09/catholics-gay-priests-grindr-data-bishops.

Bouk, Dan. *Democracy's Data*. New York: Macmillan, 2022. https://us.macmillan.com/books/9780374602543/democracysdata.

Buckley, Dave, Mark Durkee, Joe Near, and David Darais. "Privacy Attacks in Federated Learning." *Responsible Technology Adoption Unit* (blog), January 24, 2024. https://rtau.blog.gov.uk/2024/01/24/privacy-attacks-in-federated-learning.

Bun, Mark, and Thomas Steinke. "Concentrated Differential Privacy: Simplifications, Extensions, and Lower Bounds." In *Proceedings, Part I, of the 14th International Conference on Theory of Cryptography—Volume 9985*, 635–658. Berlin: Springer, 2016. https://doi.org/10.1007/978-3-662-53641-4_24.

Cohen, Aloni, and Kobbi Nissim. "Towards Formalizing the GDPR's Notion of Singling Out." *Proceedings of the National Academy of Sciences* 117, no. 15 (April 14, 2020): 8344–8352. https://doi.org/10.1073/pnas.1914598117.

"Court Case Tracker: Alabama v. U.S. Dep't of Commerce." Brennan Center for Justice. Accessed March 2, 2024. https://www.brennancenter.org/our-work/court-cases/alabama-v-us-dept-commerce.

Dalenius, Tore. "Towards a Methodology for Statistical Disclosure Control." *Statistisk Tidskrift* (*Statistical Review*)15 (1977): 429–444. https://hdl.handle.net/1813/111303.

Differential Privacy Team. "Learning with Privacy at Scale." Apple Machine Learning Research, December 2017. https://machinelearning.apple.com/research/learning-with-privacy-at-scale.

Diffie, W., and M. Hellman. "New Directions in Cryptography." *IEEE Transactions on Information Theory* 22, no. 6 (November 1976): 644–654. https://doi.org/10.1109/TIT.1976.1055638.

Ding, Bolin, Janardhan Kulkarni, and Sergey Yekhanin. "Collecting Telemetry Data Privately." In *Proceedings of the 31st International Conference on Neural Information Processing Systems*, 3574–3583. Red Hook, NY: Curran Associates, Inc., 2017. https://dl.acm.org/doi/10.5555/3294996.3295115.

Dinur, Irit, and Kobbi Nissim. "Revealing Information while Preserving Privacy." In *Proceedings of the Twenty-Second ACM SIGMOD-SIGACT-SIGART Symposium on Principles of Database Systems*, 202–210. New York: Association for Computing Machinery, 2003. https://doi.org/10.1145/773153.773173.

Dodis, Yevgeniy, Leonid Reyzin, and Adam Smith. "Fuzzy Extractors: How to Generate Strong Keys from Biometrics and Other Noisy Data." In *Advances in Cryptology—EUROCRYPT 2004*, edited by Christian Cachin, Jan L. Camenisch, and Rafail Ostrovsky, 523–540. Berlin: Springer, 2004. https://doi.org/10.1007/978-3-540-24676-3_31.

Dove, Iain. "Applying Differential Privacy Protection to ONS Mortality Data, Pilot Study." Office for National Statistics, August 20, 2021. https://www.ons.gov.uk/peoplepopulationandcommunity/birthsdeathsandmarriages/deaths/methodologies/applyingdifferentialprivacyprotectiontoonsmortalitydatapilotstudy.

Du, Minxin, Xiang Yue, Sherman S. M. Chow, Tianhao Wang, Chenyu Huang, and Huan Sun. "DP-Forward: Fine-Tuning and Inference on Language Models with Differential Privacy in Forward Pass." In *Proceedings of the 2023 ACM SIGSAC Conference on Computer and Communications Security*, 2665–2679. New York: Association for Computing Machinery, 2023. https://doi.org/10.1145/3576915.3616592.

Dwork, Cynthia. "Differential Privacy." In *Automata, Languages, and Programming*, edited by Michele Bugliesi, Bart Preneel, Vladimiro Sassone, and Ingo Wegener, 1–12. Berlin: Springer, 2006. https://doi.org/10.1007/11787006_1.

Dwork, Cynthia. "Reminiscences." *Journal of Privacy and Confidentiality* 8, no. 1 (December 2, 2018). https://doi.org/10.29012/jpc.702.

Dwork, Cynthia, Vitaly Feldman, Moritz Hardt, Toniann Pitassi, Omer Reingold, and Aaron Roth. "The Reusable Holdout: Preserving Validity in Adaptive Data Analysis." *Science* 349, no. 6248 (August 7, 2015): 636–638. https://doi.org/10.1126/science.aaa9375.

Dwork, Cynthia, Frank McSherry, Kobbi Nissim, and Adam Smith. "Calibrating Noise to Sensitivity in Private Data Analysis." In *Theory of Cryptography*, edited by Shai Halevi and Tal Rabin, 265–284. Berlin: Springer, 2006. https://doi.org/10.1007/11681878_14.

Dwork, Cynthia, and Moni Naor. "On the Difficulties of Disclosure Prevention in Statistical Databases or the Case for Differential Privacy." *Journal of Privacy and Confidentiality* 2, no. 1 (September 1, 2010). https://doi.org/10.29012/jpc.v2i1.585.

Dwork, Cynthia, Moni Naor, Toniann Pitassi, and Guy N. Rothblum. "Differential Privacy under Continual Observation." In *Proceedings of the Forty-Second ACM Symposium on Theory of Computing*, 715–724. New York: Association for Computing Machinery, 2010. https://doi.org/10.1145/1806689.1806787.

Dwork, Cynthia, and Kobbi Nissim. "Privacy-Preserving Datamining on Vertically Partitioned Databases." Microsoft, 2004, 528–544. https://www.microsoft.com/en-us/research/publication/privacy-preserving-datamining-on-vertically-partitioned-databases.

Dwork, Cynthia, and Aaron Roth. "The Algorithmic Foundations of Differential Privacy." *Foundations and Trends® in Theoretical Computer Science* 9, no. 3–4 (2013): 211–407. https://doi.org/10.1561/0400000042.

Dwork, Cynthia, and Guy N. Rothblum. "Concentrated Differential Privacy." arXiv, March 16, 2016. https://doi.org/10.48550/arXiv.1603.01887.

Dwork, Cynthia, and Jonathan Ullman. "The Fienberg Problem: How to Allow Human Interactive Data Analysis in the Age of Differential Privacy." *Journal of Privacy and Confidentiality* 8, no. 1 (December 21, 2018). https://doi.org/10.29012/jpc.687.

Ebadi, Hamid, David Sands, and Gerardo Schneider. "Differential Privacy: Now It's Getting Personal." *ACM SIGPLAN Notices* 50, no. 1 (January 14, 2015): 69–81. https://doi.org/10.1145/2775051.2677005.

Erlingsson, Úlfar, Vitaly Feldman, Ilya Mironov, Ananth Raghunathan, Kunal Talwar, and Abhradeep Thakurta. "Amplification by Shuffling: From Local to

Central Differential Privacy via Anonymity." In *Proceedings of the Thirtieth Annual ACM-SIAM Symposium on Discrete Algorithms*, 2468–2479. Philadelphia: Society for Industrial and Applied Mathematics, 2019.

Erlingsson, Úlfar, Vasyl Pihur, and Aleksandra Korolova. "RAPPOR: Randomized Aggregatable Privacy-Preserving Ordinal Response." In *Proceedings of the 2014 ACM SIGSAC Conference on Computer and Communications Security*, 1054–1067. New York: Association for Computing Machinery, 2014. https://doi.org/10.1145/2660267.2660348.

Federal Committee on Statistical Methodology. *Statistical Policy Working Paper 22: Report on the Statistical Disclosure Limitation Methodology*. Washington, DC: Office of Management and Budget, December 2005. https://www.hhs.gov/sites/default/files/spwp22.pdf.

Fellegi, I. P. "On the Question of Statistical Confidentiality." *Journal of the American Statistical Association* 67, no. 337 (March 1, 1972): 7–18. https://doi.org/10.1080/01621459.1972.10481199.

Fertility and Family Statistics Branch. "Frequently Asked Questions about Same-Sex Couple Households." US Census Bureau, August 2013. https://www2.census.gov/topics/families/same-sex-couples/faq/sscplfactsheet-final.pdf.

Finley, Klint. "It's Time to Encrypt the Entire Internet." *WIRED*, April 17, 2014. https://www.wired.com/2014/04/https.

Foote, Andrew David, Ashwin Machanavajjhala, and Kevin McKinney. "Releasing Earnings Distributions Using Differential Privacy: Disclosure Avoidance System for Post-Secondary Employment Outcomes (PSEO)." *Journal of Privacy and Confidentiality* 9, no. 2 (October 18, 2019). https://doi.org/10.29012/jpc.722.

Fredrikson, Matthew, Eric Lantz, Somesh Jha, Simon Lin, David Page, and Thomas Ristenpart. "Privacy in Pharmacogenetics: An End-to-End Case Study of Personalized Warfarin Dosing," 23rd USENIX Security Symposium, August 20–22, 2014, 17–32. https://www.usenix.org/conference/usenixsecurity14/technical-sessions/presentation/fredrikson_matthew.

Garfinkel, Simson. *De-Identification of Personal Information*. Gaithersburg, MD: National Institute of Standards and Technology, October 22, 2015. https://doi.org/10.6028/NIST.IR.8053.

Garfinkel, Simson L. "Modernizing Disclosure Avoidance: Report on the 2020 Disclosure Avoidance Subsystem as Implemented for the 2018 End-to-End

Test." Paper presented at the 2017 Census Scientific Advisory Committee, Suitland, MD, September 15, 2017. https://www2.census.gov/cac/sac/meetings/2017-09/garfinkel-modernizing-disclosure-avoidance.pdf.

Garfinkel, Simson, Joseph Near, Aref N. Dajani, Phyllis Singer, and Barbara Guttman. *De-Identifying Government Datasets: Techniques and Governance*. Gaithersburg, MD: National Institute of Standards and Technology, September 14, 2023. https://doi.org/10.6028/NIST.SP.800-188.

Greenberg, Andy. "How One of Apple's Key Privacy Safeguards Falls Short." *WIRED*, September 15, 2017. https://www.wired.com/story/apple-differential-privacy-shortcomings.

Hedrick, Shelly. "Census Protections Evolve Continuously to Address Emerging Threats." United States Census Bureau, February 3, 2020. https://www.census.gov/library/stories/2020/02/through-the-decades-how-the-census-bureau-protects-your-privacy.html.

Hill, Kashmir. "'God View': Uber Allegedly Stalked Users for Party-Goers' Viewing Pleasure (Updated)." *Forbes*, October 3, 2014. https://www.forbes.com/sites/kashmirhill/2014/10/03/god-view-uber-allegedly-stalked-users-for-party-goers-viewing-pleasure.

Hoffman, Lance J., and William F. Miller. "Getting a Personal Dossier from a Statistical Data Bank." *Datamation* 16, no. 5 (1970): 74–75.

"Information as Confidential; Exception." Title 13—Census, August 31, 1954. https://www.govinfo.gov/content/pkg/USCODE-2022-title13/pdf/USCODE-2022-title13-chap1-subchapI-sec9.pdf.

Ito, Shinsuke, Takayuki Miura, Hiroto Akatsuka, and Masayuki Terada. "Differential Privacy and Its Applicability for Official Statistics in Japan—A Comparative Study Using Small Area Data from the Japanese Population Census." In *Privacy in Statistical Databases: UNESCO Chair in Data Privacy, International Conference, PSD 2020, Tarragona, Spain, September 23–25, 2020, Proceedings*, 337–352. Berlin: Springer, 2020. https://doi.org/10.1007/978-3-030-57521-2_24.

Jain, Palak, Sofya Raskhodnikova, Satchit Sivakumar, and Adam Smith. "The Price of Differential Privacy under Continual Observation." In *Proceedings of the 40th International Conference on Machine Learning*, 202:14654–14678. Honolulu: JMLR, 2023.

Kasiviswanathan, Shiva P., and Adam Smith. "On the 'Semantics' of Differential Privacy: A Bayesian Formulation." *Journal of Privacy and Confidentiality* 6, no. 1 (June 1, 2014). https://doi.org/10.29012/jpc.v6i1.634.

Kennickell, Arthur, and Julia Lane. "Measuring the Impact of Data Protection Techniques on Data Utility: Evidence from the Survey of Consumer Finances." In *Privacy in Statistical Databases*, edited by Josep Domingo-Ferrer and Luisa Franconi, 4302:291–303. Berlin: Springer, 2006. https://doi.org/10.1007/11930242_25.

Kifer, Daniel, John M. Abowd, Robert Ashmead, Ryan Cumings-Menon, Philip Leclerc, Ashwin Machanavajjhala, William Sexton, and Pavel Zhuravlev. "Bayesian and Frequentist Semantics for Common Variations of Differential Privacy: Applications to the 2020 Census." arXiv, September 7, 2022. https://doi.org/10.48550/arXiv.2209.03310.

Kohnfelder, Loren M. "Towards a Practical Public-Key Cryptosystem." Master's thesis, Massachusetts Institute of Technology, 1978. https://dspace.mit.edu/handle/1721.1/15993.

Kraus, Rebecca. "Statistical Déjà Vu: The National Data Center Proposal of 1965 and Its Descendants." *Journal of Privacy and Confidentiality* 5, no. 1 (August 1, 2013). https://doi.org/10.29012/jpc.v5i1.624.

Kühtreiber, Patrick, Viktoriya Pak, and Delphine Reinhardt. "Replication: The Effect of Differential Privacy Communication on German Users' Comprehension and Data Sharing Attitudes." In *Proceedings of the Eighteenth USENIX Conference on Usable Privacy and Security*, 117–134. Berkeley, CA: USENIX Association, 2022.

Machanavajjhala, Ashwin, Daniel Kifer, John Abowd, Johannes Gehrke, and Lars Vilhuber. "Privacy: Theory Meets Practice on the Map." In *2008 IEEE 24th International Conference on Data Engineering*, 277–286. New York: IEEE, 2008. https://doi.org/10.1109/ICDE.2008.4497436.

Mathieu, Edouard, Hannah Ritchie, Lucas Rodés-Guirao, Cameron Appel, Charlie Giattino, Joe Hasell, Bobbie Macdonald, et al. "COVID-19: Google Mobility Trends." Our World in Data, March 5, 2020. https://ourworldindata.org/covid-google-mobility-trends.

McSherry, Frank. "Differential Privacy for Dummies." *GitHub* (blog), February 3, 2016. https://github.com/frankmcsherry/blog/blob/master/posts/2016-02-03.md.

McSherry, Frank D. "Privacy Integrated Queries: An Extensible Platform for Privacy-Preserving Data Analysis." In *Proceedings of the 2009 ACM SIGMOD International Conference on Management of Data*, 19–30. New York: Association for Computing Machinery, 2009. https://doi.org/10.1145/1559845.1559850.

McSherry, Frank. "Response to Criticism of 'Fools Gold' Article." *GitHub* (blog), May 19, 2016. https://github.com/frankmcsherry/blog/blob/master /posts/2016-05-19.md.

McSherry, Frank, and Kunal Talwar. "Mechanism Design via Differential Privacy." In *48th Annual IEEE Symposium on Foundations of Computer Science (FOCS'07)*, 94–103. New York: IEEE, 2007. https://doi.org/10.1109/FOCS .2007.66.

Medina, Caroline, and Lindsay Mahowald. "Key Issues Facing People with Intersex Traits." Center for American Progress, October 26, 2021. https://www .americanprogress.org/article/key-issues-facing-people-intersex-traits.

Miller, Arthur. "Personal Privacy in the Computer Age: The Challenge of a New Technology in an Information-Oriented Society." *Michigan Law Review* 67, no. 6 (April 1, 1969): 1089–1247. https://repository.law.umich.edu/mlr/vol67 /iss6/2.

Minkel, J. R. "Confirmed: The U.S. Census Bureau Gave Up Names of Japanese-Americans in WW II." *Scientific American*, March 30, 2007. https://www.scien tificamerican.com/article/confirmed-the-us-census-b.

Mironov, Ilya. "On Significance of the Least Significant Bits for Differential Privacy." In *Proceedings of the 2012 ACM Conference on Computer and Communications Security*, 650–661. New York: Association for Computing Machinery, 2012. https://doi.org/10.1145/2382196.2382264.

Mironov, Ilya. "Rényi Differential Privacy." In *30th Computer Security Foundations Symposium*, 263–275. New York: IEEE, 2017. https://doi.org/10.1109 /CSF.2017.11.

Narayanan, Arvin, and Vitaly Shmatikov. "Robust De-Anonymization of Large Sparse Datasets." In *2008 IEEE Symposium on Security and Privacy*. New York: IEEE, 2008. https://ieeexplore.ieee.org/document/4531148.

Nasr, Milad, Jamie Hayes, Thomas Steinke, Borja Balle, Florian Tramèr, Matthew Jagielski, Nicholas Carlini, and Andreas Terzis. "Tight Auditing of Differentially Private Machine Learning," 1631–1648. 32nd USENIX Security

Symposium, 2023. https://www.usenix.org/conference/usenixsecurity23/presentation/nasr.

National Cancer Institute Press Office. "Three Breast Cancer Gene Alterations in Jewish Community." US Department of Health and Human Services, May 20, 1997.

Near, Joseph P., and Chiké Abuah. *Programming Differential Privacy*. 2021. https://programming-dp.com.

Near, Joseph P., and Xi He. "Differential Privacy for Databases." *Foundations and Trends® in Databases* 11, no. 2 (July 21, 2021): 109–225. https://doi.org/10.1561/1900000066.

"1940 Census Records. National Archives, September 8, 2016. https://www.archives.gov/research/census/1940.

Nunes, Gabriel H., Mário S. Alvim, and Annabelle McIver. "A Formal Quantitative Study of Privacy in the Publication of Official Educational Censuses in Brazil." In *Anais Do Concurso de Teses e Dissertações (CTD)*, 61–70. Anais do XXXV Concurso de Teses e Dissertações, SBC, 2022. https://doi.org/10.5753/ctd.2022.223158.

Oberski, Daniel L., and Frauke Kreuter. "Differential Privacy and Social Science: An Urgent Puzzle." *Harvard Data Science Review* 2, no. 1 (January 31, 2020). https://doi.org/10.1162/99608f92.63a22079.

Office for Civil Rights. "Guidance regarding Methods for De-Identification of Protected Health Information in Accordance with the Health Insurance Portability and Accountability Act (HIPAA) Privacy Rule." US Department of Health and Human Services, October 25, 2022. https://www.hhs.gov/hipaa/for-professionals/privacy/special-topics/de-identification/index.html.

"Pillar Investigates: USCCB Gen Sec Burrill Resigns after Sexual Misconduct Allegations," July 20, 2021. https://www.pillarcatholic.com/p/pillar-investigates-usccb-gen-sec.

"Recital 26: Not Applicable to Anonymous Data." GDPR, April 27, 2016. https://gdpr.eu/recital-26-not-applicable-to-anonymous-data.

Rivest, Ronald L., Adi Shamir, and Len Adleman. "A Method for Obtaining Digital Signatures and Public-Key Cryptosystems." DSpace@MIT, April 1977. https://dspace.mit.edu/handle/1721.1/148910.

Rodríguez, Rolando. "Confidentiality for American Community Survey Respondents." Paper presented at the 2023 ACS Data Users Conference, May 16, 2023. https://acsdatacommunity.prb.org/p/conferences.

Romer, Paul M. "Mathiness in the Theory of Economic Growth." *American Economic Review* 105, no. 5 (May 2015): 89–93. https://doi.org/10.1257/aer.p20151066.

Schneider, Mike. "Census Releases Guidelines for Controversial Privacy Tool." AP News, June 9, 2021. https://apnews.com/article/business-census-2020-55519b7534bd8d61028020d79854e909.

"1790 Census Records." National Archives, October 28, 2020. https://www.archives.gov/research/census/1790.

Shankland, Stephen. "Google Begins Blurring Faces in Street View." *CNET*, May 13, 2008. https://www.cnet.com/culture/google-begins-blurring-faces-in-street-view.

Shokri, Reza, Marco Stronati, Congzheng Song, and Vitaly Shmatikov. "Membership Inference Attacks against Machine Learning Models." In *2017 IEEE Symposium on Security and Privacy*, 3–18. In New York: IEEE, 2017. https://doi.org/10.1109/SP.2017.41.Smid, Miles. "Development of the Advanced Encryption Standard." *Journal of Research of the National Institute of Standards and Technology* 126, no. 126024 (2021). https://doi.org/10.6028/jres.126.024.

Solove, Daniel. "A Taxonomy of Privacy." *University of Pennsylvania Law Review* 154, no. 3 (January 1, 2006): 477–560. https://scholarship.law.upenn.edu/penn_law_review/vol154/iss3/1.

Song, Shuang, Kamalika Chaudhuri, and Anand D. Sarwate. "Stochastic Gradient Descent with Differentially Private Updates." In *2013 IEEE Global Conference on Signal and Information Processing*, 245–248. New York, IEEE, 2013. https://doi.org/10.1109/GlobalSIP.2013.6736861.

Sprenger, Polly. "Sun on Privacy: 'Get Over It.'" *WIRED*, January 26, 1999. https://www.wired.com/1999/01/sun-on-privacy-get-over-it.

Sweeney, Latanya. "Only You, Your Doctor, and Many Others May Know." *Technology Science*. Accessed October 21, 2023. https://techscience.org/a/2015092903.

Taft, Howard William. "By the President of the United States of America: A Proclamation." March 15, 1910. https://www.census.gov/history/img/procla mation1910-artifact.jpg.

Tezapsidis, Katie. "Uber Releases Open Source Project for Differential Privacy." *Uber Privacy and Security* (blog), July 13, 2017. https://medium.com /uber-security-privacy/differential-privacy-open-source-7892c82c42b6.

"2020 Census Operational Plan v4.0." Suitland, MD: US Census Bureau, December 2018. https://www2.census.gov/programs-surveys/decennial/2020 /program-management/planning-docs/2020-oper-plan4.pdf.

"Uber Settles FTC Allegations That It Made Deceptive Privacy and Data Security Claims." Federal Trade Commission, August 15, 2017. https://www.ftc.gov /news-events/news/press-releases/2017/08/uber-settles-ftc-allegations-it -made-deceptive-privacy-data-security-claims.

United States Bureau of the Census. *U.S. Census of Population: 1960: Availability of Published and Unpublished Data*. Washington, DC: Department of Commerce, 1964.

US Census Bureau. "Authorizing the First Census—The Significance of Population Data." US Census Bureau. Accessed September 9, 2023. https://www .census.gov/programs-surveys/sis/activities/history/first-census.html.

US Census Bureau. *Disclosure Avoidance for the 2020 Census: An Introduction*. Washington, DC: US Government Printing Office, November 2021. https:// www2.census.gov/library/publications/decennial/2020/2020-census-disclo sure-avoidance-handbook.pdf.

US Census Bureau. "Monograph: Census Confidentiality and Privacy: 1790–2002." United States Census Bureau. Accessed September 10, 2023. https:// www.census.gov/library/publications/2003/comm/monograph-confidential ity-privacy.html.

US Department of Education. *Technical Documents: College Scorecard Data by Field of Study*. February 2024. https://collegescorecard.ed.gov/assets/Field OfStudyDataDocumentation.pdf.

Warner, Stanley L. "Randomized Response: A Survey Technique for Eliminating Evasive Answer Bias." *Journal of the American Statistical Association*, March 1, 1965. https://www.tandfonline.com/doi/abs/10.1080/01621459.1965.104 80775.

Warren, Samuel D., and Louis D. Brandeis. "The Right to Privacy." *Harvard Law Review* 4, no. 5 (1890): 193–220. https://doi.org/10.2307/1321160.

Welch, Chris. "Uber Will Pay $20,000 Fine in Settlement over 'God View' Tracking." *Verge*, January 6, 2016. https://www.theverge.com/2016/1/6/10726004/uber-god-mode-settlement-fine.

"Welcome." OpenDP, 2024. https://docs.opendp.org.

Wright, Carroll D., and William C. Hunt. *History and Growth of the United States Census: 1790–1890*. Washington, DC: Government Printing Office, 1900.

FURTHER READING

Dwork, Cynthia, and Aaron Roth, *The Algorithmic Foundations of Differential Privacy* (Norwell, MA: Now Publishers Inc., August 11, 2014). https://www.cis.upenn.edu/~aaroth/Papers/privacybook.pdf.

Garfinkel, Simson, Joseph Near, Aref N. Dajani, Phyllis Singer, and Barbara Guttman. *De-Identifying Government Datasets: Techniques and Governance*. Gaithersburg, MD: National Institute of Standards and Technology, September 14, 2023. https://doi.org/10.6028/NIST.SP.800-188.

Near, Joseph P., and Chiké Abuah. *Programming Differential Privacy*. 2021. https://programming-dp.com.

"Welcome." OpenDP, 2024. https://docs.opendp.org.

Wood, Alexandra, Micah Altman, Kobbi Nissim, and Salil Vadhan, "Designing Access with Differential Privacy." In *Handbook on Using Administrative Data for Research and Evidence-Based Policy*, edited by Shawn Cole, Iqbal Dhaliwal, Anja Sautmann, and Lars Vilhuber, chapter 6. Cambridge, MA: J-PAL, 2021. https://admindatahandbook.mit.edu/book/latest/diffpriv.html.

INDEX

Note: Page numbers in italics indicate illustrations; *t* denotes tables.

SIMSON GARFINKEL works on projects at the intersection of artificial intelligence, privacy, data management, and digital forensics. Garfinkel holds seven US patents, and has published dozens of research articles on topics including computer security, privacy, and digital forensics.

Garfinkel received three bachelor of science degrees from MIT in 1987, a master of science in journalism from Columbia University in 1988, and a PhD in computer science from MIT in 2005.